Mastering AI-Driven Business Transformation

Empowering Business Leaders to Drive Measurable AI-Enabled Transformation

Steve Twist

Published by Twist Tech, Sydney, Australia.

Twist Tech is an imprint of TBC Academy Pty Ltd.

First edition: 2026

ISBN (paperback): 978-1-7644839-0-2

Trademarks

All trademarks, service marks, and logos referenced in this book are the property of their respective owners. The use of any trademark does not imply endorsement by the trademark holder.

Disclaimer

This book is provided for educational and informational purposes only. It does not constitute professional advice of any kind, including but not limited to legal, financial, compliance, cybersecurity, human resources, or operational advice. You should consult qualified professionals before making decisions or taking actions based on the content of this book.

While the publisher and author have made reasonable efforts to ensure the accuracy and completeness of the information provided, no representations or warranties are made regarding the accuracy, reliability, or suitability of the content for any particular purpose. The content is provided "as is." Results may vary based on organizational context, data quality, regulatory requirements, market conditions, and execution.

To the fullest extent permitted by law, the publisher and author disclaim any liability for any loss, damage, or disruption arising from the use of, or reliance on, this book, including direct, indirect, incidental, consequential, or special damages, even if advised of the possibility of such damages.

Examples and scenarios are provided for illustrative purposes only. Any names, organizations, or situations are fictionalized unless otherwise noted.

AI and Data Responsibility Notice

Artificial intelligence systems may produce inaccurate, incomplete, biased, or otherwise unreliable outputs. Any AI-related methods, tools, or practices described in this book should be implemented with appropriate governance, security controls, human oversight, and validation.

Readers are responsible for ensuring compliance with all applicable laws, regulations, contractual obligations, and industry standards, including privacy and data protection requirements. Do not use confidential, personal, or regulated data with AI systems unless you have appropriate authorization and safeguards in place.

Company and Product References

References to specific products, services, frameworks, organizations, or vendors are for informational purposes only and do not constitute endorsement, sponsorship, or recommendation.

Permissions

For permissions or licensing inquiries: contact@twisttech.net

Preface

A few years ago, I sat in a conference room with a leadership team doing what many organizations do when AI enters the conversation. They were eager to "do something with AI" and were trying to decide where to start.

The CIO had a backlog of aging systems. The COO was focused on reducing cycle times and improving operational reliability. The business unit leader wanted growth, not another technology program that produced dashboards nobody used. I was the enterprise architect in the room, quietly taking notes as the discussion bounced between big ambitions and real constraints.

At one point, I put my pen down and said, "We do not have an AI problem. We have an architecture and alignment problem."

The room went quiet, not because anyone was surprised, but because it finally named what everyone had been feeling. They were not short on ideas. They were short on a shared way to turn ideas into outcomes.

That is the core reason I wrote this book.

AI is powerful, but AI on its own is not a strategy. Enterprise architecture is powerful, but not when it is treated as an internal IT exercise. When you bring them together, something changes. AI-driven enterprise architecture, the discipline of aligning strategy, operating model, and technology so AI becomes a repeatable capability, not a series of isolated pilots, becomes a practical way to connect strategy to execution, and execution to measurable outcomes.

In this book, AI-driven enterprise architecture means designing your operating model, data, systems, and governance so AI can be deployed safely, scaled reliably, and tied directly to business outcomes.

You also get a common language that helps business and IT teams stop talking past each other and start building together.

Who This Book Is For

If you are a business leader or executive, you will find guidance on building the business case, aligning IT and business goals, leading change, and communicating value, risk, and ROI with clarity.

If you are an enterprise architect or IT professional, you will find practical approaches to designing scalable AI-enabled architectures, integrating with legacy environments, strengthening governance and security, and translating technical decisions into business terms that leaders can act on.

How to Use This Book

You can read straight through or use it as a playbook.

- Start with the fundamentals, then move into the business case and alignment.
- Use the architecture, change, governance, and measurement chapters as your execution toolkit.
- Return to the case studies and future trends when you need patterns, inspiration, and a forward view.

A Note on Responsible Use

AI can accelerate decisions, but it can also accelerate mistakes. That is why this book treats ethics, governance, and security as core design requirements, not optional add-ons. The goal is not to slow you down. The goal is to help you move fast with confidence, with appropriate oversight, and with respect for privacy, fairness, and accountability.

One Last Thought Before You Begin

If you have ever felt stuck between a bold ambition and a complex reality, you are in the right place. This book is designed to help you connect strategy to architecture, architecture to execution, and execution to measurable outcomes.

You will find frameworks, tools, and case studies, but you will also find a consistent message: transformation is achievable when you treat AI-driven enterprise architecture as a leadership discipline, not a technology trend.

By the end of this book, you will be able to lead AI initiatives from business case to architecture to measurable outcomes.

Let's get started.

> The hardest part of AI is rarely the model, it's the operating model.

Table of Contents

Table of Figures

Key Terms at a Glance

AI foundations

- Artificial Intelligence (AI)
- Narrow AI
- General AI
- Superintelligent AI
- Machine Learning (ML)
- Deep Learning
- Artificial Neural Network
- Natural Language Processing (NLP)
- Predictive Analytics
- Automation
- Model Retraining
- Human-in-the-loop

Data and decision intelligence

- Data Governance
- Data Quality
- Data Pipeline
- Data Steward
- Metadata
- Master Data
- Privacy-by-design
- Consent Management
- Audit Trail
- Explainability

Enterprise architecture and modern platforms

- Enterprise Architecture (EA)
- AI-Driven Enterprise Architecture
- Operating Model
- Capability Model
- Target Architecture
- Architecture Principles
- Reference Architecture
- TOGAF
- Architecture Development Method (ADM)
- Agile Enterprise Architecture
- Model-Driven Architecture (MDA)
- Cloud-native Architecture
- Microservices
- APIs
- Distributed Data
- Load Balancing
- Observability

Strategy, value, and execution

- Digital Transformation
- Business Case
- Value Realization
- Total Cost of Ownership (TCO)
- Return on Investment (ROI)
- Cost-benefit Analysis
- Scenario Planning
- Benchmarking
- Key Performance Indicators (KPIs)
- SMART Objectives

Change, teams, and leadership

- Change Management
- Kotter's 8-Step Model
- Prosci ADKAR Model
- Change Champions
- Cross-functional Teams
- Psychological Safety
- Collaboration
- Stakeholder Buy-in
- Communication Plan
- Continuous Learning
- AI Literacy

Ethics, governance, and security

- AI Governance
- Responsible AI
- Fairness
- Bias
- Bias Audit
- Transparency
- Accountability
- Compliance
- GDPR
- CCPA
- HIPAA
- Adversarial Attack
- Role-based Access Control (RBAC)
- Penetration Testing

Future trends

- Internet of Things (IoT)
- Edge Computing
- 5G
- Blockchain
- Digital Assistants
- AI Orchestration

Introduction

A familiar Monday morning

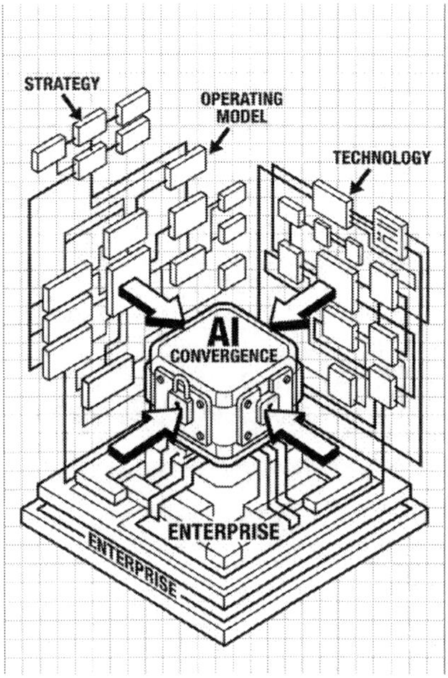

Picture this. You open your laptop on a Monday morning and the first thing you see is another headline about a market leader losing ground to a faster, smarter competitor. The story is always slightly different, but the pattern is the same.

A new player enters with a sharper understanding of customers, faster decisions, and a product experience that keeps improving in real time. Meanwhile, established organizations scramble to respond. Not because they lack talent or ambition, but because their systems, operating model, and decision cycles were built for a different pace.

That is the environment we are operating in now. Business models evolve in months, not years. Customer expectations reset constantly. And artificial intelligence is no longer a side experiment. It is becoming the engine behind speed, insight, and adaptability.

So here is the practical truth: if you are leading transformation, AI is not something you "add on." It is something you deliberately design for.

Why this book exists

This book is about AI-driven enterprise architecture, and the reason that phrase matters is simple.

AI creates value when it is connected to how your organization actually works: how decisions get made, how processes run end to end, how data moves, how risk is managed, and how change gets delivered. Enterprise architecture is the discipline that connects those pieces.

When AI and enterprise architecture work together, you get more than better technology. You get a clearer line of sight from strategy to execution. You build systems that can learn, adapt, and scale. And most importantly, you make it easier for business and IT to move in the same direction without losing speed or control.

If you are a business leader, that means you can translate AI investment into measurable outcomes: growth, efficiency, resilience, and better customer experiences.

If you are an enterprise architect or IT leader, it means you can design AI capabilities that are secure, governed, reusable, and aligned with what the business is trying to accomplish, rather than a collection of disconnected pilots.

What "AI-driven enterprise architecture" looks like in practice

Let's make this tangible.

It can look like customer insights that show up in time to influence decisions, not weeks later in a slide deck. It can look like processes that are increasingly automated and self-correcting, freeing teams to focus on higher-value work. It can look like leaders having confidence in decisions because the data is trusted, the models are governed, and the organization knows when to rely on AI and when not to.

The real goal is not to sprinkle AI across the company and hope something sticks. The goal is to build an enterprise that can repeatedly turn AI capabilities into business value, without creating chaos, risk, or dependency on a few heroic individuals.

The obstacles you are probably facing

If you have lived through transformation, you already know the hardest part is rarely the algorithm.

The hard part is the environment around it.

You might be navigating competing priorities and a patchwork of legacy systems. You might be dealing with fragmented data, unclear ownership, or inconsistent definitions that make "insight" a moving target. You might be worried about security, privacy, regulatory exposure, and the reputational damage that comes from irresponsible AI use. And you might be feeling the reality of change fatigue, where people nod in meetings but resist in practice because they have seen too many initiatives come and go.

None of this makes you behind. It makes you honest about what it takes to do this well.

This book does not treat AI transformation as effortless or inevitable. It treats it as a leadership and architecture challenge that can be solved with the right approach, the right sequencing, and a clear connection to outcomes.

What you will get from this book

You will not find vague inspiration here. You will find a practical guide designed to help you lead and implement AI-enabled transformation in a way that is measurable, responsible, and scalable.

You will learn how to:

- Connect AI initiatives to business outcomes and measurable value
- Align IT and business strategy so you can move faster with less friction
- Build the architectural foundations that make AI repeatable, not fragile
- Strengthen enterprise architecture maturity to support modern AI capabilities
- Design governance that balances innovation, risk management, and trust
- Lead organizational change with less resistance and more adoption
- Communicate clearly with stakeholders to secure buy-in, funding, and momentum
- Build cross-functional teams that can deliver complex AI programs

- Future-proof your organization by adapting to emerging trends without chasing hype

Throughout the book, I will translate between executive priorities and technical realities, because successful transformation depends on both.

Business leaders need a clear path from investment to outcome. Architects and delivery teams need concrete patterns, sequencing, and guardrails that hold up under enterprise constraints.

My perspective and what I am drawing from

My perspective comes from years at the intersection of strategy, technology implementation, and enterprise leadership. I have seen what happens when transformation is approached as a slide deck exercise, and I have also seen what is possible when strategy, architecture, and delivery are aligned.

In developing this book, I drew from established methods and practical experience. You will see the discipline of enterprise architecture, including TOGAF principles where helpful, combined with modern delivery practices such as agile, design thinking, and change management. You will also see responsible AI considerations woven in, because governance and ethics are not separate workstreams. They are part of how you build trust and sustain impact.

The examples and stories are designed to be recognizable across industries. The goal is not to give you a "perfect blueprint." The goal is to give you patterns that work in the real world, and the judgment to adapt them.

How the book is organized

We will move in a clear sequence, because sequencing is where many AI programs succeed or fail.

First, we will establish the foundations by clarifying what AI is, what it is not, and how enterprise architecture helps you make it real inside complex organizations.

Next, we will build the business case. You will learn how to articulate value, align investments to strategy, and secure executive commitment without overpromising.

Then we will get into design and integration. We will cover principles and patterns for scalable, adaptable AI-enabled architectures, including how to think about data, platforms, integration, and reuse.

After that, we will focus on change leadership. Because even the best architecture fails if adoption fails. We will address how to lead organizational shifts, build capability, and maintain momentum when the initial excitement wears off.

We will also tackle governance, risk, and responsible AI. You will learn how to set policies and decision mechanisms that enable innovation while protecting the organization, customers, and employees.

Finally, we will focus on measurement and sustainability. You will learn how to track progress, demonstrate impact, iterate intelligently, and embed AI-driven improvements into ongoing operations. We will also look ahead to emerging trends, so you can plan with confidence without chasing every new headline.

Each chapter includes practical tools, checklists, and case examples, including missteps. You will learn what works, what fails, and why.

Who this book is for

If you are a business leader responsible for transformation, this book is written to help you turn AI ambition into measurable results. It will help you connect technology decisions to business outcomes, lead change with less resistance, and set your organization up for sustained advantage.

If you are an enterprise architect or IT professional implementing AI solutions, this book is written to help you design architectures that are scalable and aligned, improve maturity over time, and bridge the gap between technical teams and business stakeholders.

If you are both, you are in exactly the right place.

A quick commitment before we begin

One promise I will make to you: we will stay grounded. We will aim for clarity over complexity, outcomes over activity, and progress over perfection.

AI-driven enterprise architecture is not a technology project. It is a business strategy expressed through systems, operating models, and decisions.

Done well, it gives you a repeatable way to convert emerging AI capabilities into sustainable value, while balancing innovation with risk, compliance, and operational excellence.

So, let's begin. Bring your curiosity, but also bring your skepticism. That combination is what separates AI theater from AI transformation.

> AI changes the speed of decisions.
> Architecture protects the quality of them

CHAPTER 1

Understanding AI and Enterprise Architecture

A late afternoon that tells you everything you need to know

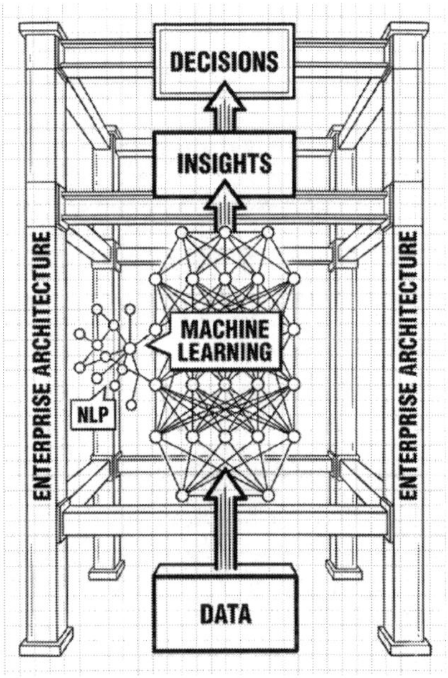

Late one afternoon, a customer service manager watched her team struggle to keep up with a sudden surge of inquiries.

Calls piled up, emails went unanswered, and customer frustration rose with every minute. At the same time, the IT director was being asked a different question: "Which systems do we fix first?"

The budget was tight, the backlog was growing, and every team had a different opinion.

Across the organization, leaders felt surrounded by data but starved for insight. Everyone sensed that technology should be helping more than it was, yet it was not obvious where to start or how to pull the pieces together.

If that scene feels familiar, you are not alone. It is a snapshot of a deeper tension: businesses want speed, clarity, and adaptability, while their technology landscape often reflects years of layered decisions, competing priorities, and uneven modernization. That is why this chapter starts with fundamentals. Not academic fundamentals, but practical ones. The kind that let you lead better conversations, make better decisions, and build momentum without guessing.

By the end of this chapter, you will understand what AI really is, what enterprise architecture really does, and why the combination creates leverage for transformation.

Fundamentals of AI and Enterprise Architecture

AI in plain English

Artificial intelligence is not a single tool, and it is not magic. It is a family of techniques that let machines perform tasks that normally require human judgment, such as recognizing patterns, learning from examples, understanding language, and recommending actions.

You have already used AI today if you interacted with a chatbot, received a fraud alert from your bank, or got a recommended product that was surprisingly relevant.

The enterprise question is not "Does AI exist?" The enterprise question is "How do we use AI intentionally to create measurable outcomes?"

A helpful way to think about AI as a leader is this:

- **AI turns data into predictions, classifications, and decisions at scale.**
- **AI automates some work and augments other work.**
- **AI's value depends on the quality of data, the design of the process, and the trust people have in the results.**

Why AI matters to enterprises

AI matters because it changes how quickly organizations can sense what is happening and respond.

In a customer support environment, a well-designed AI assistant can handle routine questions, surface relevant knowledge articles, and help agents resolve complex cases faster.

In retail, AI can forecast demand and reduce out-of-stock events. In banking, AI can flag unusual transactions and strengthen fraud detection.

The thread connecting these examples is not the novelty of the technology, it's the business effect: faster decisions, better use of talent, fewer errors, and more responsiveness.

The three levels of AI you should keep straight

Conversations about AI often get muddy because people blend different concepts together. Keeping these categories separate helps you stay grounded:

- **Narrow AI:** Systems designed to do a specific task well, such as invoice processing, document classification, forecasting, or summarizing customer conversations. Most practical enterprise use cases today are narrow AI.
- **General AI:** A theoretical capability with human-like flexibility across many domains. It does not exist today in the way people often imagine it.
- **Superintelligent AI:** A further theoretical concept where machine intelligence surpasses human intelligence in broad and significant ways.

Most enterprise transformation decisions today are about narrow AI. When people debate long-term existential risk or sweeping societal impact, they are often discussing general or superintelligent AI.

As a business leader or architect, you will make better choices when you keep your focus on what is real, available, and governable.

The building blocks that power AI

You do not need to become a data scientist, but you do need to know the components well enough to ask the right questions.

Machine learning

Machine learning learns patterns from historical examples to make predictions or classifications in new situations. A retailer uses it to forecast demand. A finance team uses it to identify transactions that look unusual.

Deep learning

Deep learning is a subset of machine learning that can process large volumes of unstructured data such as images, audio, and text. In a factory, it can detect defects from camera feeds faster and more consistently than manual inspection.

Natural language processing (NLP)

NLP enables systems to work with human language, written or spoken. It can route emails, extract intent from customer messages, summarize call transcripts, and support more natural chatbot interactions.

Data as the foundation

AI is only as good as the data beneath it. If data is incomplete, inconsistent, or untrusted, AI outcomes degrade quickly. That is why data governance is not an "IT hygiene" initiative. It is a business enabler.

Good governance sets standards for how data is defined, cleaned, labeled, protected, and shared. It supports compliance with privacy and security obligations while making data usable for analytics and AI.

One insurer that struggled with claims delays improved outcomes by correcting data errors and organizing customer histories before introducing a claims-processing AI model.

The result was fewer mistakes, faster turnaround, and improved customer satisfaction.

Here is the key point: **AI is powerful, but it needs an environment where it can be trusted, scaled, and sustained.**

That is where enterprise architecture enters the story.

What enterprise architecture really is

Enterprise architecture is the discipline of designing and governing how an organization's business processes, information, applications, and technology infrastructure work together to achieve strategy.

If that sounds broad, it is, and that is the point. Enterprise architecture connects decisions that are often made in isolation:

- Strategy and operating model decisions
- Process design and accountability

- Data and integration design
- Application portfolios and platform choices
- Risk, security, and compliance controls
- Delivery sequencing and dependency management

When enterprise architecture is working well, it does not slow transformation down. It reduces rework, prevents fragmentation, and creates clarity about what should be standardized, what can be flexible, and what needs to evolve first.

For leaders, enterprise architecture offers a way to answer questions like:

- What capabilities do we need to win in our market?
- Which systems and data domains are critical to those capabilities?
- Where are we taking on hidden risk or unnecessary complexity?
- What is the smartest modernization sequence based on value and feasibility?

Where AI and enterprise architecture intersect

AI amplifies both the opportunity and the risk of enterprise change. That is why the intersection matters.

Better decisions, not just better dashboards

AI can help architectural teams see what is happening across the enterprise by analyzing streams of business and technical data from applications, customer interactions, infrastructure telemetry, and operational workflows. This improves the ability to identify failure trends, performance bottlenecks, and operational pain points.

Predictive analytics can also anticipate issues before they become outages or major business disruptions (Tweed, 2025).

A retailer preparing for seasonal demand spikes, for example, can use AI to simulate where data pipelines or backend services are likely to strain, then target investments where they matter most.

This shifts decisions from "best guess" to "evidence-based prioritization."

Linking technology roadmaps to business outcomes

When AI-driven insights are connected to business priorities, modernization becomes more strategic. Instead of spreading resources across too many initiatives, teams can focus on the areas most likely to improve customer experience, reduce cost, or increase resilience.

A large organization running a legacy platform might use AI to continuously monitor reliability, user experience signals, and maintenance cost. The models can help identify which modules yield the greatest return if modernized first.

The benefit is straightforward: IT spend becomes more defensible, risk declines, and outcomes become easier to measure (Schrage & Kiron, 2025).

Automating the work that traps architects

Many architecture teams spend a surprising amount of time on documentation, compliance checks, and audit preparation. AI can automate portions of that work by scanning logs, design artifacts, and operational metadata to flag compliance gaps or outdated documentation (Tweed, 2025).

Consider a financial services firm using an AI documentation assistant. When a service is redeployed or an integration changes, the assistant updates configuration references, adds change notes, and refreshes diagrams in the enterprise repository.

Architects spend less time chasing accuracy and more time guiding design choices, managing dependencies, and aligning with business priorities.

That is not just efficiency. It is capacity creation for higher-value work.

Architectures that adapt in near real time

Modern architectures need to respond to change in demand, location, and usage patterns. AI can help by learning from operational and business data, then adjusting configurations and resources accordingly.

In digital entertainment, traffic spikes can occur based on releases or viral trends. AI-driven orchestration platforms can learn patterns and scale capacity in the regions affected, then scale down when traffic stabilizes. The organization improves customer experience while controlling cost (Tweed, 2025).

Faster collaboration between business and IT

AI assistants are increasingly useful for knowledge sharing, especially in large organizations where critical information is spread across tools, teams, and documents.

Business analysts and architects can ask questions and receive immediate context: data definitions, integration dependencies, documentation references, and known risks.

Imagine an international manufacturer running an architecture change workshop with engineering, operations, and compliance stakeholders. An AI-enabled collaboration platform captures proposals, maps them to existing patterns and business cases, and highlights potential risks or overlaps. This increases participation, shortens decision cycles, and strengthens buy-in for implementation (Schrage & Kiron, 2025).

The bigger point is this: **enterprise architecture creates the structure for alignment, and AI accelerates insight, automation, and adaptability inside that structure.**

AI's impact on business transformation

When AI is integrated into an enterprise architecture intentionally, it stops being a series of disconnected experiments. It becomes a transformation engine that improves operations, customer experiences, and innovation capacity across the organization.

Operational efficiency through automation and prediction

AI is especially effective in high-volume workflows that contain repeatable patterns. In customer service, AI can handle routine inquiries and support agents in complex cases.

In manufacturing and logistics, machine learning models can monitor equipment signals to predict maintenance needs before failures occur.

Predictive maintenance reduces downtime and extends asset life (Gibson, 2024).

The practical business effect is that teams spend less time on repetitive work and more time on higher-value analysis, design, and decision-making.

Over time, those gains become a competitive advantage because the organization can reinvest savings into innovation and capability building.

Customer experiences that feel personal at scale

AI-driven personalization has become a core expectation in many markets. By analyzing behavior patterns, purchase history, and context, organizations can tailor experiences in near real time.

NLP can identify intent and sentiment, helping service teams respond with the right tone and the right resolution path.

Predictive analytics extends this by forecasting customer needs, enabling better timing for offers, retention actions, and support interventions.

Streaming platforms, for example, use these techniques to recommend content based on user habits, increasing satisfaction and engagement (Gibson, 2024).

When executed well, customers spend less time searching, receive fewer irrelevant offers, and experience faster problem resolution.

Trust grows when the experience consistently feels helpful rather than intrusive.

New products, new services, and faster experimentation

AI can lower the cost and risk of testing new business models by improving how quickly organizations learn. An insurance provider might use AI to support usage-based pricing, while retail organizations can adjust promotions and replenishment decisions based on real-time signals.

These capabilities allow teams to test ideas, gather feedback, and iterate quickly (Ali et al., 2024).

This strengthens a culture of experimentation because learning loops tighten. Organizations become better at responding to market shifts without betting everything on a single long-cycle initiative.

The human side: adoption, capability, and trust

Technology does not transform organizations. People do.

Resistance is common, especially when employees worry about job impact or feel overwhelmed by constant change.

Organizations reduce friction when they invest in capability building, transparent communication, and practical support.

Upskilling programs, hands-on workshops, and clear guidance help teams build confidence. Leaders who communicate the purpose, the trade-offs, and the expected benefits create trust.

Pairing experienced users with less familiar colleagues accelerates adoption. Cultivating curiosity and making space for safe experimentation strengthens long-term outcomes (Ali et al., 2024).

This is how AI-driven transformation becomes sustainable rather than a one-time push.

Wrapping up

In this chapter, we grounded the conversation in what is real and useful.

AI is a set of techniques that turns data into predictions, automation, and decision support. Enterprise architecture is the discipline that connects strategy, operations, and technology so change can be delivered coherently. When you integrate the two, you gain leverage: clearer prioritization, stronger alignment between business and IT, more repeatable execution, and the ability to adapt faster with less chaos.

In the next chapter, we will build on these foundations and move toward the business conversation leaders care about most: how to translate AI ambition into a clear business case, measurable outcomes, and a roadmap your organization can actually deliver.

AI is a capability.
Enterprise Architecture is how it becomes reliable

CHAPTER 2

The Business Case for AI-Driven Transformation

The budget meeting that changes the conversation

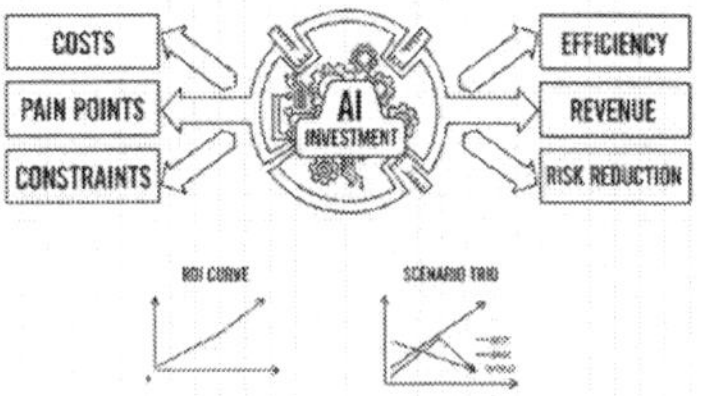

A transformation lead walks into a budget review with a simple request: funding for an AI initiative that promises faster service, lower cost, and better customer experiences.

The CFO asks, "How much value, by when, and how will we prove it?"

The COO asks, "What changes for my teams on day one, and what breaks if adoption is slow?"

The CIO asks, "How does this fit with our architecture roadmap, data constraints, and security obligations?"

Three leaders. Three valid questions. One common theme: nobody wants AI as a shiny upgrade. They want a business case that is clear, measurable, and credible.

That is what this chapter is about. We will move from AI excitement to disciplined decision-making, so you can prioritize the right opportunities, quantify value and risk, and communicate in a way that creates alignment across business and IT.

Why AI is a business strategy, not a technology purchase

Artificial intelligence is already reshaping how organizations compete. The advantage does not come from using AI once. It comes from building the capability to repeatedly turn data and automation into measurable outcomes.

Organizations that treat AI as "a project" often end up with pilots that never scale. Organizations that treat AI as "a strategy" make different choices:

- They target outcomes, not features.
- They design for adoption, not just accuracy.
- They invest in foundations like data, governance, and operating model changes.
- They measure value continuously, not only at the end.

Your business case is the mechanism that forces these choices into the open.

Identifying AI business opportunities

Start where friction is expensive

The best AI opportunities usually hide in plain sight: repetitive work, slow decision cycles, and areas where variability creates cost or risk.

A structured assessment beats "gut feel" because it helps you see patterns across departments.

A practical starting framework looks like this:

1. **Map the workflow** end to end (not just the piece in one team).
2. **Find measurable pain points** (time, cost, error, delay, rework).
3. **Assess AI fit** (volume, repetition, data availability, decision rules).
4. **Estimate value potential** (financial and strategic).
5. **Confirm feasibility constraints** (data, integration, governance, change readiness).

Process mapping often reveals where manual work is silently draining capacity, like invoice processing, onboarding steps, compliance checks, or reporting cycles (Takyar, 2023).

Once you see the workflow, you can identify specific bottlenecks with clear metrics: where delays accumulate, where errors occur, and where teams spend time correcting preventable issues.

Example: A retail company processes thousands of invoices manually each week. The work is repetitive and error prone. That combination is a strong signal for document automation using AI, replacing slow data entry with faster, more consistent extraction and validation.

Look for "high-volume, high repeat" first

Not every problem needs AI. Many problems can be solved with better process design, clearer ownership, or basic automation. AI becomes attractive where three conditions overlap:

- High volume of transactions or decisions
- High repetition or recognizable patterns

- Clear success criteria (even if the path to success varies)

That is why early wins often show up in customer service triage, document processing, forecasting, anomaly detection, knowledge search, and scheduling support (Takyar, 2023).

Use your own data to validate where value exists

Once you have candidate opportunities, validate them with real operational data. Look at response times, conversion rates, defect rates, backlog trends, and customer complaint themes.

These signals often point to where automation or decision support will pay off.

Analytics-driven identification reduces risk because it replaces "we think this will help" with "we can see where performance breaks" (Bhardwaj, 2025).

Predictive analytics pushes you from reactive to proactive. Instead of responding after churn happens, you anticipate it. Instead of firefighting outages, you predict capacity strain.

The business case gets stronger when you can show that the AI use case changes the timing of decisions, not just the speed of execution.

> Your best AI use case is the one you can prove.

Make opportunity discovery cross-functional on purpose

Many of the highest-value use cases sit between departments. That is exactly where they get missed.

A simple pattern that works well is a structured workshop with process owners, architects, data leaders, and risk or compliance partners.

The goal is to identify end-to-end opportunities, not isolated team improvements.

Cross-functional collaboration also increases buy-in because the people who will live with the change help shape it. Many organizations formalize this by appointing "AI champions" within business units who help guide pilots, provide feedback, and keep adoption moving.

This approach is especially effective for use cases like dynamic pricing, procurement risk scoring, and supply chain optimization, where the inputs and impacts span multiple functions (Takyar, 2023).

Benchmark externally, but do not copy blindly

External benchmarking helps you calibrate what is possible and what "good" looks like. Leaders study digital leaders in their sector and compare internal metrics against industry performance, adoption patterns, and published outcomes.

Benchmarking can help you prioritize proven opportunities like demand forecasting, visual inspection, and customer personalization (Bhardwaj, 2025; Takyar, 2023).

The key is to **translate external examples into your context**, constraints, and strategic priorities.

AI Opportunity Scorecard (quick template)

Use this to shortlist opportunities before anyone gets attached to a pet idea:

- **Business value potential:** low, medium, high
- **Time to impact** under 3 months, 3 to 9 months, 9 months plus
- **Data readiness:** poor, partial, strong
- **Integration complexity:** low, medium, high
- **Risk and compliance sensitivity:** low, medium, high
- **Adoption effort:** low, medium, high
- **Strategic alignment:** weak, moderate, strong

Conducting cost-benefit analysis for AI projects

Identifying opportunities is the fun part. Funding decisions happen when value becomes measurable and assumptions become explicit.

Step 1: Establish baselines that everyone accepts

Before you estimate benefits, capture today's reality:

- Average handling time
- Cycle time and throughput
- Error rates and rework
- Cost per transaction
- Customer satisfaction and retention signals
- Risk metrics (incidents, compliance findings, losses)

Baseline agreement matters because it prevents debates later about whether the project "really worked."

Step 2: Capture total cost, including the parts that surprise people

AI ROI fails most often because costs were understated. Build a complete cost view that includes:

- Data preparation and governance work
- Integration and platform changes
- Security and privacy controls
- Model monitoring and maintenance
- Training and change management
- Ongoing licensing and infrastructure costs
- Temporary productivity dips during adoption

Many ROI models emphasize the need to include these categories from the start (DX, 2025; Proving ROI - Measuring the Business Value of Enterprise AI - Agility at Scale, 2025).

Step 3: Quantify benefits in two categories

Tangible benefits

These show up in the financial model:

- Reduced labor hours for repetitive tasks
- Lower error and rework costs
- Higher throughput and faster cycle times
- Reduced downtime and operational losses
- Improved conversion rates and revenue capture

Intangible benefits

These are real, but need disciplined framing:

- Higher customer satisfaction and loyalty
- Better decision speed and quality
- Lower risk exposure and stronger compliance posture
- Brand reputation and market credibility
- Increased employee capacity for higher-value work

Example ROI logic (simplified):

An AI service chatbot costs $750,000 for launch, training, and integration, plus $150,000 per year for maintenance and licensing. It reduces workload equivalent to 12 full-time roles at $80,000 each, or $960,000 per year.

Year one net savings are modest once you include ramp-up and adoption, but years two and three improve as the model stabilizes and the operating model adapts.

This is also where intangible gains like retention lift can materially strengthen the case (Proving ROI - Measuring the Business Value of Enterprise AI - Agility at Scale, 2025).

Step 4: Do scenario planning to keep the case honest

You want three scenarios:

- **Best case:** strong adoption, expected accuracy, smooth integration
- **Most likely:** expected bumps, moderate adoption curve, manageable issues
- **Worst case:** slow adoption, underestimated maintenance, integration delays

Scenario planning forces a useful question: "What assumptions must be true for this ROI to hold?" It also helps leaders plan mitigation early instead of being surprised later.

Here is a simple way to visualize it without overcomplicating:

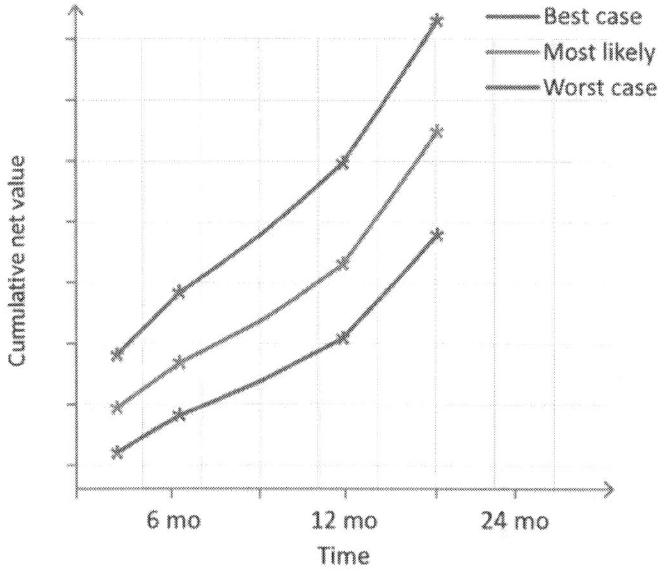

Figure 1 - ROI Scenario Graph

Step 5: Allocate costs and benefits fairly across departments

Enterprise AI rarely lives in one cost center. If the investment is shared, the business case should reflect shared accountability.

Allocate costs by who benefits and who uses the capability. For example, IT may fund platform integration, while operations fund training and adoption work in their teams. Data platform upgrades may be shared across multiple programs. Clear allocation reduces political friction and improves transparency (DX, 2025).

Step 6: Commit to value tracking after launch

The business case is not a one-time document. **Value is proven over time**.

Set formal reviews at 6, 12, and 24 months. Track both financial metrics and operating model outcomes. Some organizations discover that early savings undercount the real value because freed-up capacity is redirected to high-impact work, such as strategy, innovation, or risk reduction (Proving ROI - Measuring the Business Value of Enterprise AI - Agility at Scale, 2025).

Establishing objectives and communicating AI value

A strong financial model still fails if objectives are vague and communication is unclear. People cannot execute what they cannot explain.

Turn ambition into SMART outcomes

Vague goals like "use AI to innovate" are not measurable, and they do not drive alignment. The SMART framework gives structure:

- **Specific:** What exactly changes?
- **Measurable:** Which metric proves progress?
- **Achievable:** Do we have the capability and time?
- **Relevant:** Does this support strategy?
- **Time-bound:** When will we deliver measurable impact?

Example: "Reduce average customer support response time by 30 percent within six months through chatbot deployment." This kind of goal sharpens priorities and makes progress visible (Harris, 2023).

Choose KPIs that match the value story

KPIs translate objectives into management signals. They should cover both performance and adoption, because adoption is where value is either realized or lost.

Common AI transformation KPIs include:

- Automation rate (percentage of cases handled by AI)
- Cycle time reduction
- Error rate reduction
- Model precision and false positives
- Cost per transaction
- Revenue lift or conversion improvement
- Customer satisfaction and retention signals
- User adoption and usage frequency

A retail chain might target "increase online conversion by 15 percent this quarter" and track recommendation-driven sales and ecommerce revenue.

A financial services firm targeting "reduce fraud by 25 percent" might track detected fraud cases and model false-positive rate (Proving ROI - Measuring the Business Value of Enterprise AI - Agility at Scale, 2025; Harris, 2023).

Benchmark to stay grounded and improve continuously

Benchmarking helps you understand whether your progress is meaningful. Compare against internal history and external peers. It keeps teams honest and prevents both complacency and unrealistic expectations (Proving ROI - Measuring the Business Value of Enterprise AI - Agility at Scale, 2025).

Build feedback loops that keep the transformation on track

AI systems and workflows evolve. Your goals and approach should evolve too.

Use:

- Staff surveys to identify friction and adoption barriers
- Stakeholder interviews to detect shifting priorities
- Dashboards to track usage and value signals

Feedback loops help teams refine scope, improve relevance, and increase adoption (Harris, 2023). They also reduce the risk of scaling a solution that looks good on paper but fails in the lived experience of users.

Communicate value differently to different stakeholders

One message does not work for everyone.

- **C-suite:** value, risk, timeline, competitive advantage
- **Business unit leaders:** operational outcomes, customer impact, team implications
- **IT and architects:** feasibility, integration, security, delivery sequencing
- **Finance:** cost transparency, assumptions, sensitivity, payback timeline

Make value tangible with simple visuals and short stories: hours saved, cycle time improvement, customer satisfaction trends, error reductions. Regular updates build trust and prevent the "AI mystery project" effect (Proving ROI - Measuring the Business Value of Enterprise AI - Agility at Scale, 2025; Harris, 2023).

Executive Lens: the questions that make or break the business case

If you are sponsoring AI-driven transformation, these are the questions to insist on:

- **What business outcome are we buying, and how will we measure it?**
- **What must change in process, roles, and decisions to realize the value?**
- **What risks are introduced, and how are they mitigated?**
- **What is the scaling plan if the pilot succeeds?**
- **What is the kill criteria if it does not?**
- **How does this align to enterprise priorities and funding realities?**

A credible business case includes both ambition and constraints. It shows the path to value and the discipline to stop or adjust when reality differs from the plan.

> If you can't measure it, you can't scale it.

Architecture and Delivery Lens: making the business case executable

A business case that ignores architecture usually becomes a pilot that cannot scale.

Architects and delivery leaders should pressure-test:

- **Data readiness:** Is data accessible, governed, and fit for use?
- **Integration:** Where does the AI capability sit in the flow of work?
- **Operationalization:** Who monitors model drift, performance, and failure modes?
- **Security and privacy:** What controls are required from day one?
- **Reuse:** Can we build capabilities once and apply them across use cases?
- **Sequencing:** What prerequisites must be delivered before value shows up?

When the architecture lens is integrated early, the business case becomes more realistic, and the delivery path becomes clearer.

Common pitfalls and how to avoid them

Pitfall: Starting with the coolest use case instead of the most valuable one

Countermove: score opportunities against measurable outcomes and feasibility.

Pitfall: Underestimating data preparation and governance effort

Countermove: treat data readiness as a first-class investment and include it in ROI (DX, 2025).

Pitfall: Assuming adoption is automatic

Countermove: budget for change management, training, and workflow redesign.

Pitfall: Measuring only technical performance

Countermove: track adoption and business outcomes, not just model accuracy.

Pitfall: Declaring victory after a pilot

Countermove: define a scaling plan and operating model from the start.

Signals you are on track

- Leaders can describe the value case in one minute without handwaving.
- Baselines are agreed and progress is visible on a small set of KPIs.
- Business and IT share accountability for outcomes, not just delivery.
- Adoption is rising steadily, not flat after launch.
- The organization is building reusable capabilities, not one-off models.
- Governance and risk controls are enabling progress, not blocking it.

Case snapshots

Success pattern: scaling an operational win

A customer operations team starts with a focused use case: AI-supported triage and routing for inbound requests. The business case is grounded in response time, customer satisfaction, and staff capacity.

The team invests early in data quality and workflow integration, then measures outcomes monthly. As results stabilize, the organization scales the approach to additional channels and regions.

KPI tracking and ROI reviews remain active after launch, strengthening confidence to expand (Proving ROI - Measuring the Business Value of Enterprise AI - Agility at Scale, 2025).

Failure pattern: a pilot with no path to value

A team launches an AI prototype that looks impressive in demos but is not integrated into daily workflows. Data ownership is unclear, adoption support is minimal, and success metrics are vague.

Six months later, the model still exists, but usage is low and the business cannot explain measurable impact. Funding dries up, and stakeholders become sceptical of the next proposal.

The core issue is not the technology. It is the absence of a business case that connected outcomes, operating model change, and delivery realities.

What I would do Monday morning

1. Build a shortlist of 10 to 15 candidate AI opportunities using workflow mapping.
2. Score them with the AI Opportunity Scorecard and select the top 3.
3. Define baselines for each: cost, time, quality, risk, customer impact.
4. Draft a one-page business case per opportunity:
   - Outcome
   - KPIs
   - Cost categories
   - Risks and mitigations
   - Best, most likely, worst scenarios
5. Identify the cross-functional owners:
   - Business process owner
   - Data owner
   - Architecture lead
   - Security and risk partner
   - Change lead
6. Decide what "scale" means before you pilot:
   - Target scope
   - Reuse plan
   - Operating model for monitoring and support
7. Schedule value reviews at 6, 12, and 24 months and assign accountability.

Key takeaway

AI-driven transformation becomes real when the business case is disciplined: clear opportunities, honest costs, measurable outcomes, and a communication plan that builds trust across business and IT.

That discipline turns AI from a hopeful experiment into a repeatable advantage.

Bridge to the next chapter

Now that you can build a credible business case, the next question is unavoidable: how do you design the architecture and delivery approach, so these initiatives scale safely and sustainably?

In the next chapter, we will move from "why and what" to "how," focusing on the architectural principles and integration patterns that make AI transformation repeatable across the enterprise.

> ROI is not a spreadsheet outcome.
> It's an operating outcome.

CHAPTER 3

Strategic Alignment of IT and Business Goals

The meeting where everyone agrees and nothing moves

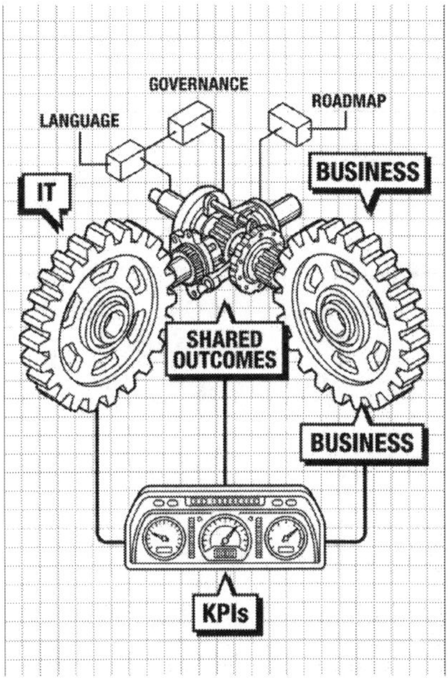

A senior leader ends a steering committee meeting with a confident line: "We are aligned."

Then the next week, the business asks for faster delivery, IT asks for more time to reduce risk, Finance asks for a tighter scope, Security asks for stronger controls, and Product asks for new features.

Nobody is being unreasonable, but progress slows anyway.

If you have lived through this, you know the frustrating truth: alignment is not a statement. It is a system.

It shows up in how decisions are made, how priorities are set, how trade-offs are handled, and how accountability works across boundaries.

AI raises the stakes. When business and IT are not aligned, AI becomes a set of disconnected pilots, inconsistent data definitions, and fragile deployments that do not scale. When they are aligned, AI becomes an accelerator for strategy.

This chapter gives you practical ways to build that alignment. You will learn how to bridge business and IT conversations, form integrated teams that deliver, and develop leadership behaviors that keep alignment alive under real pressure.

Why strategic alignment matters more in an AI-driven enterprise

Strategic alignment means that business goals, IT priorities, architecture decisions, funding, and delivery plans reinforce each other. It means the organization can answer one fundamental question consistently:

How does this technology effort advance the business strategy, and how will we measure it?

Without alignment, AI initiatives often fail in predictable ways:

- The business wants outcomes, but projects deliver tools.
- IT builds models, but workflows do not change.
- Data issues surface late and derail timelines.
- Risk and compliance are treated as afterthoughts, then block scaling.
- Teams disagree on what success means.

Alignment reduces these failure modes by creating shared language, shared decision rights, and shared accountability.

Alignment is not a meeting.
It's a shared decision system.

Bridging IT and business for effective AI alignment

Make alignment a cadence, not an event

Alignment is not created in a quarterly planning session. It is created through regular, structured interaction.

Weekly or biweekly cross-functional check-ins help surface what is changing: market pressures, operational pain points, platform constraints, delivery risks, and upcoming launches. A monthly strategy review can connect near-term roadmaps to long-term priorities. The goal is simple: **reduce surprises**.

These conversations work when both sides feel safe to say what is true. Business leaders should be able to say, "This is the outcome we need." IT leaders should be able to say, "Here is the constraint, the risk, and the option that gets us there fastest."

Build a shared vocabulary that reduces friction

Misalignment often starts with language. Business leaders speak in outcomes and timing. IT leaders speak in feasibility and dependencies. Both are correct, but translation is required.

A practical tactic is a shared glossary in a collaboration space that defines common terms in plain language: APIs, data products, model drift, technical debt, and governance.

This is not busywork. It reduces meeting time, improves clarity, and prevents decisions based on misunderstandings.

Use shared dashboards to make progress visible

Visibility reduces distrust. When teams rely on informal updates, assumptions grow.

Shared dashboards provide a single view of:

- Outcomes and KPIs
- Delivery milestones
- Key risks and dependencies
- Resource allocation
- Adoption and usage signals

The value is not the dashboard itself. The value is the shared reality it creates.

Use workshops to build empathy and accelerate decisions

Cross-functional workshops work best when they are designed to solve a real problem, not just "collaborate."

A product launch delay, a capacity bottleneck, a data quality issue, or a prioritization conflict are all strong workshop topics. Build mixed teams of business, engineering, operations, security, and architecture. Give them clear constraints and a shared definition of success.

Role-reversal exercises can deepen empathy. A business leader spends time in a technical support simulation. An architect responds to rapidly shifting customer requirements.

The point is not performance. The point is understanding. Teams move faster when they respect each other's constraints and pressures.

Tie collaboration to shared outcomes

Collaboration improves when success is defined together. For example, a team working on online purchasing performance can agree on shared metrics like:

- Order completion time
- Error rate during checkout
- Customer satisfaction after purchase
- Percentage of orders completed without support intervention

Publish these measures and review them together. Celebrate improvements publicly. When the organization sees that success belongs to both business and IT, alignment stops being a slogan and becomes behavior.

Make executive sponsorship visible and specific

Even strong teams struggle without leadership signal. Executive sponsorship matters most when it is visible and practical:

- Sponsors attend key decision meetings
- Sponsors reinforce shared goals when conflict appears
- Sponsors protect teams from constant priority churn
- Sponsors reward cross-functional wins

When a bottleneck arises between IT and finance, for example, a sponsor can ask the right questions and keep the discussion anchored on outcomes rather than positions.

The sponsor does not need to dictate. The sponsor needs to ensure the organization stays aligned on what matters.

Keep momentum with frequent reinforcement

Alignment fades when wins are invisible. Leaders should recognize small successes, not just large milestones: a prototype delivered early, a deployment stabilized, a workflow improved, a risk mitigated. This builds energy and signals that collaboration is valued.

As teams experience these wins, they naturally ask a bigger question: what would we accomplish if we worked as one integrated group rather than two coordinated groups?

That is where the next section begins.

Building integrated teams for AI project success

Putting people in the same room is not integration. Integration means shared objectives, clear roles, and a working rhythm that supports learning.

Define the essential roles for an AI team

AI projects require both technical expertise and business context. A balanced team typically includes:

- Data scientists to explore data and develop models
- Machine learning engineers to build, deploy, and scale models
- Data engineers to create and maintain pipelines
- Enterprise or solution architects to ensure integration and architectural coherence
- Business analysts or product owners to connect work to business outcomes
- Domain experts who understand the process and exceptions

- Security and risk partners to guide controls and compliance
- Project or delivery leaders to manage execution and dependencies (Team, 2024)

The names can vary, but the functions matter.

Assess capability gaps early and plan for them

AI teams often fail because the organization assumes the skill set exists. A practical approach is to assess capabilities at the start:

- Technical fluency: data engineering, model deployment, monitoring
- Domain fluency: process knowledge, customer behavior, regulatory context
- Product and change fluency: adoption, workflow design, training

Where gaps exist, leaders can use targeted upskilling, hiring, or time-bound external support (Team, 2024).

The key is to plan it before delivery is blocked.

Clarify responsibilities to reduce confusion

Diverse teams move fast only when responsibilities are clear. A simple responsibility map prevents overlap and missed handoffs.

For example:

- Data engineers own pipeline reliability and data quality controls
- Data scientists own feature development and model experimentation

- Machine learning engineers own production deployment and monitoring
- Architects own integration patterns, standards, and platform fit
- Product owners own outcome definitions, prioritization, and stakeholder communication (Team, 2024)

Clarity accelerates delivery because teams stop negotiating roles mid-project.

Build a culture that rewards shared success

Culture becomes the engine of integrated delivery. Leaders shape culture through what they reward.

Encourage open dialogue in standups and reviews. Recognize team milestones publicly, and make it safe to raise concerns early. This is how you prevent late-stage surprises and build trust (Business, 2022; Team, 2024).

A practical habit is to anchor every sprint review to two questions:

- What value did we deliver toward the business outcome?
- What did we learn that changes the plan?

This keeps the team focused on outcomes and learning, not just tasks.

Create a shared vision that is repeated and used

Integrated teams deliver better when they rally behind a clear mission. Leaders should repeat the mission often and connect daily work to it.

A practical tool is a simple "north star statement" visible in dashboards and working sessions, such as:

"Reduce customer issue resolution time by 30 percent while maintaining compliance and improving customer satisfaction."

The team should also have input into how the statement evolves. Shared ownership increases commitment (Business, 2022).

Use agile practices to keep alignment continuous

Agile practices support integrated teams because they create short feedback loops:

- Sprints with clear outcomes
- Regular sprint reviews with business stakeholders
- Retrospectives focused on removing friction
- Daily standups to surface blockers early (Team, 2024)

If a retrospective reveals bottlenecks in model deployment, the team can adjust immediately. This is how integrated teams stay responsive rather than rigid.

Essential leadership skills for AI transformation

Strategic alignment lives or dies in leadership behavior. Tools and structures help, but leaders set the tone for how decisions are made and how conflict is resolved.

Visionary thinking

Alignment starts with a clear story of where the organization is going and why AI matters to that direction. Leaders who explain the purpose in practical terms help teams connect short-term projects to long-term value.

A strong vision is both bold and usable. It should make trade-offs easier, not harder. It should clarify which outcomes matter most and which distractions to avoid (DILAN, 2025).

Visionary leaders also create space for teams to bring insights about emerging trends and competitor signals. That curiosity fuels innovation and reduces the risk of being surprised by market shifts (Navigating the Future of Leadership in Business Landscapes: Key Trends for 2025 and Beyond, 2025).

Emotional intelligence

AI transformation creates uncertainty. People worry about roles, skills, and stability. Leaders with emotional intelligence recognize that resistance often comes from fear or fatigue, not stubbornness.

Practical behaviors that build trust include:

- Holding open debriefs after launches that welcome honest feedback
- Recognizing effort while addressing problems directly
- Asking open-ended questions and listening without defensiveness
- Admitting uncertainty when it exists and sharing how decisions will be made (DILAN, 2025)

When leaders model humility and curiosity, teams feel safer raising issues early. That improves delivery and alignment.

Decisive, data-driven decision-making

AI accelerates the pace of change, and teams need leaders who can make clear decisions without waiting for perfect information.

Decisive leaders use data, consult the right stakeholders, and commit. They also explain the reasoning, so teams understand the trade-offs. When decisions change, they share what was learned and what will be done differently.

That transparency builds confidence (DILAN, 2025; Navigating the Future of Leadership in Business Landscapes: Key Trends for 2025 and Beyond, 2025).

Inclusive leadership

AI initiatives benefit from diverse perspectives because they touch customers, employees, processes, and risk. Inclusive leaders draw out viewpoints that would otherwise be missed and use them to improve outcomes (DILAN, 2025).

Practical inclusive habits include:

- Structured forums where all roles contribute, not only the loudest voices
- Rotating facilitators in workshops
- Anonymous feedback channels for sensitive issues
- Sharing credit across departments when outcomes improve

When people feel heard and valued, alignment becomes easier because teams trust the process.

Wrapping up

Strategic alignment between IT and business is no longer optional, especially in AI-driven transformation. Alignment creates the conditions where AI investments translate into measurable outcomes instead of isolated pilots.

In this chapter, you learned how to build alignment through regular communication cadences, shared vocabulary, and transparent dashboards, and how integrated teams deliver better when roles are clear, culture rewards shared success, with agile practices keeping feedback loops active..

In the next chapter, we will take this alignment foundation and apply it to how architecture decisions are made at scale, focusing on principles and operating mechanisms that keep AI initiatives coherent, secure, and reusable across the enterprise.

> If IT and business agree on outcomes, the architecture gets easier.

CHAPTER 4

Designing AI-Driven Enterprise Architectures

The pilot that impressed everyone and still did not scale

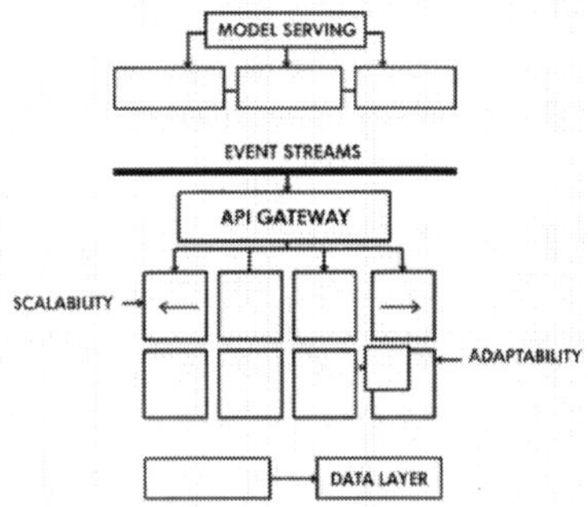

The demo was flawless.

A small team had built an AI assistant that summarized customer issues and suggested next-best actions.

In a pilot group, it saved time and earned real praise from frontline staff.

The executive sponsor walked into the steering meeting smiling, expecting the obvious next step: scale it across the organization.

Then the questions started.

"Which customer data is the model allowed to use?" "Where does it run, and how do we secure it?" "How do we keep outputs consistent across regions?" "What happens when the model starts drifting?" "Who owns the business outcome, and who owns the risk?"

The pilot was good. The enterprise was not ready.

That is the difference this chapter tackles. AI value at scale is rarely blocked by the model itself. It is blocked by architecture: data foundations, integration patterns, governance, operating mechanisms, and scalability choices that make AI repeatable, safe, and useful across the enterprise.

By the end of this chapter, you will understand how to design AI-driven enterprise architectures that align with business goals, scale responsibly, and create a platform for continuous improvement.

The core idea

An AI-driven enterprise architecture is not "AI bolted onto existing systems."

It is an architectural approach where AI capabilities are designed as part of the enterprise: connected to business processes, supported by trustworthy data, governed for risk and ethics, and engineered for resilience and scale.

A simple way to hold this in your head is:

- **AI delivers decisions and automation.**
- **Architecture delivers trust, integration, and scale.**
- **Together, they deliver measurable outcomes.**

> Architecture is the discipline that turns ambition into repeatable execution.

Frameworks and scalability for AI integration

Frameworks matter because they force discipline. They help you avoid building clever solutions that cannot survive the realities of enterprise constraints.

Framework 1: The AI-Driven Framework (outcomes first)

The AI-Driven Framework starts with business goals and works backward to AI opportunities, ensuring every AI capability has a direct link to value (Enterprise IT Services, 2025).

A practical way to apply it:

- Define strategic goals (retention, cost reduction, resilience, growth).
- Identify the decisions and workflows that drive those goals.
- Map AI opportunities to those decisions and workflows.
- Add governance gates for ethics, privacy, and accountability.
- Design for scaling from the start, not after the pilot.

Example: A retailer focused on customer retention connects behavioral analytics to marketing actions, feeding near real-time insights into personalized outreach engines rather than producing reports that sit unused (Enterprise IT Services, 2025).

Framework 2: Agile Enterprise Architecture (feedback loops and adaptability)

Agile Enterprise Architecture reduces long planning cycles by using shorter feedback loops and iterative delivery. Instead of assuming you can design the perfect architecture up front, you design for learning and adjustment.

AI projects benefit from this because requirements and data realities evolve quickly. A team can pilot a chatbot, learn from real user conversations, improve sentiment analysis, and adjust the workflow integration as needs change (Jager, 2025).

Agile EA is not "less architecture." It is architecture that stays close to delivery and value.

Framework 3: Model-Driven Architecture (clarity before implementation)

Model-Driven Architecture (MDA) emphasizes modeling business flows, data flows, and constraints before committing to a technical solution. This is especially useful when AI introduces new logic that can become hard to explain later.

In regulated environments, MDA can help surface risks early by simulating flows and identifying audit and control needs before code is written.

The discipline is helpful when you need explainability and traceability in AI-enabled decisions.

Framework 4: TOGAF, adapted for AI

TOGAF offers a structured way to design enterprise architecture, but AI changes what "good" looks like.

TOGAF becomes more powerful when you intentionally plug AI considerations into the architecture lifecycle (Jager, 2025).

Examples of AI-specific additions include:

- Capability assessments that include data readiness and AI maturity
- Evaluation criteria that include bias detection and ethical impact
- Governance checkpoints for model monitoring and accountability
- Using AI to map real workflows, not just documented workflows (Enterprise IT Services, 2025)

This keeps AI innovation aligned with enterprise standards and regulatory needs.

The architecture patterns that make AI scalable

Frameworks guide decisions, but patterns make them real. These are the architectural choices that often determine whether AI becomes a repeatable enterprise capability.

Cloud-native foundations for elastic demand

AI workloads fluctuate. Cloud-native approaches give you on-demand capacity so you can scale processing up and down with real demand.

Using this approach, a retailer can handle holiday spikes in transactions and deliver recommendations in near real time without overbuilding infrastructure for the rest of the year (Jager, 2025).

Microservices for safer change and faster experimentation

Microservices break AI-enabled capabilities into smaller components that can evolve independently allowing a bank to update fraud detection models or scoring logic one service at a time without risking disruptions across the entire system.

This supports both innovation and resilience, because change is isolated and reversible.

Load balancing and resilience for real-world peaks

AI experiences spikes too. Chatbots see surges during outages. Recommendation engines spike during campaigns. Load balancing ensures capacity is directed where it is needed most so performance remains stable and customer trust stays intact (Enterprise IT Services, 2025).

Distributed data patterns for global enterprises

AI depends on timely data. Distributed data architectures allow insight generation across geographies without severe bottlenecks or latency. This matters when decisions depend on global demand signals, supply chain shifts, or regional customer behavior.

A practical blueprint: the minimum AI architecture stack

Here is a simple way to visualize what must exist for AI to scale beyond a pilot:

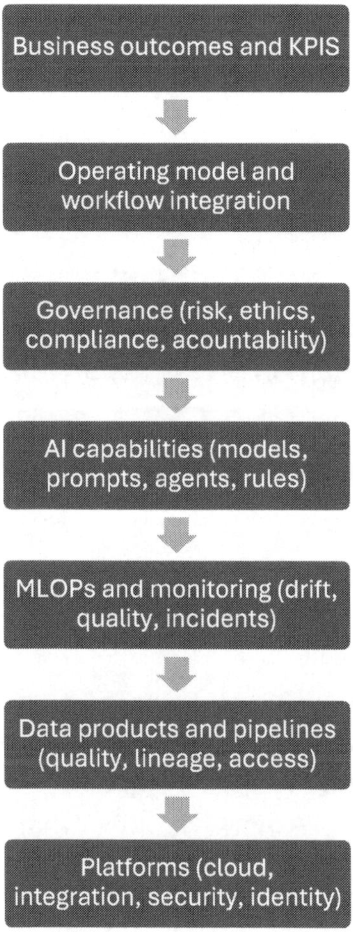

Figure 2 - AI Architecture Stack

If any layer is missing, scaling becomes fragile.

Innovation techniques through AI-enabled processes

Once the architecture foundations exist, the real payoff shows up in how work changes.

Automation that creates capacity, not just savings

AI automation is often the fastest value lever because it targets repetitive work. Document automation, as an example, reduces manual invoice entry and frees teams to analyze spending and negotiate supplier improvements.

Chatbots can handle routine support questions and triage urgent issues so humans can focus on complex cases (Gibson, 2024).

The biggest shift is not hours saved. It is the capacity created to improve processes and build new value.

Predictive analytics that change timing and decision quality

Predictive analytics turns reporting into foresight. Demand forecasting improves when models consume real-time logistics, sales, and market signals, allowing proactive inventory decisions instead of reactive firefighting (Rajagopalan, 2020).

This applies beyond the supply chain. HR can anticipate skill shortages and plan recruitment and training earlier, and Finance can improve forecasting confidence with stronger signals.

Continuous learning from customer feedback loops

AI can digest feedback from live chat, surveys, social channels, and product reviews, turning raw signals into actionable insight (Gibson, 2024). This supports faster product and service adjustments and enables personalization at scale.

The winning pattern is simple: listen continuously, act quickly, measure the result, and repeat.

Collaboration at speed with AI-enabled knowledge and coordination tools

AI-driven collaboration tools reduce the "status work" that slows teams down. Summarizers, intelligent task management, knowledge bases, and automated documentation can reduce friction across departments, especially in complex transformation programs (Rajagopalan, 2020).

When teams spend less time chasing updates, they spend more time solving problems and improving outcomes.

Case studies illustrating AI architecture design success

Real-world examples show a consistent lesson: architecture design choices drive whether AI becomes operational advantage.

Retail: inventory and demand optimization

Retailers have used AI-driven analytics that combine sales data, historical trends, and supplier dynamics to improve stocking decisions. The architectural focus is typically on end-to-end data flow integration across channels and making recommendations usable for store teams through clear dashboards.

Case studies report reductions in inventory costs and fewer stockouts when predictive models are supported by strong data integration and frequent refresh cycles (Michał, 2024; Dilmegani, 2018).

Financial services: risk assessment and compliance

Financial institutions face strict regulatory requirements, which makes explainability, auditability, and data security essential.

AI-enabled risk systems have combined machine learning with NLP for document parsing, then delivered explainable risk scores with traceable outcomes.

Case studies report faster compliance processing and improved fraud investigation speed when architectures include strong privacy safeguards and feedback loops for model improvement (Dilmegani, 2018; Michał, 2024).

Healthcare: predictive modelling with privacy at the core

Healthcare predictive modelling often depends on integrating fragmented clinical data such as notes, labs, and device metrics. Effective architectures typically include anonymization mechanisms, carefully governed access, and clinician-aligned workflows.

Case studies report reductions in readmissions when predictive models are integrated into care coordination processes rather than living as standalone analytics (Michał, 2024; Dilmegani, 2018).

Manufacturing: predictive maintenance and IoT integration

Manufacturers have used IoT sensor streams combined with cloud processing and machine learning to predict equipment failures before they cause downtime. Successful designs emphasize reliable pipelines, retraining cycles, and reporting that maintenance teams can act on quickly.

Case studies report meaningful downtime reductions when predictive maintenance is deployed as an integrated operating capability, not just a monitoring tool (Dilmegani, 2018; Michał, 2024).

Executive Lens: what leaders should insist on

If you are sponsoring AI-driven architecture decisions, focus on these questions:

- What business outcomes are we targeting, and which KPIs prove success?
- Which decisions and workflows will change, and who owns adoption?
- What data domains are critical, and who owns quality and access?
- What risks are introduced, and what controls are designed in from day one?
- How will we scale from pilot to enterprise without rebuilding everything?
- What is the operating model for monitoring, support, and continuous improvement?

Leaders who ask these questions early prevent the most common failure: exciting pilots that cannot survive enterprise reality.

Architecture and Delivery Lens: what makes AI "enterprise-ready"

Architects and delivery leaders should validate that the design includes:

- Data products with clear ownership, quality standards, and lineage
- Integration patterns that place AI into the flow of work
- Monitoring for drift, performance, and failure modes
- Security and privacy controls aligned to risk sensitivity
- A deployment pipeline that supports iteration and governance gates
- Resilience patterns like load balancing, fallbacks, and graceful degradation (Enterprise IT Services, 2025; Jager, 2025)

This is what turns AI into a reusable capability rather than a one-off system.

Common pitfalls and how to avoid them

Pitfall: Treating architecture as documentation instead of a decision system

Countermove: make architecture artifacts decision-oriented, tied to outcomes and delivery sequencing.

Pitfall: Scaling models without scaling data governance

Countermove: establish data ownership, standards, and access patterns early, then build AI on top.

Pitfall: Ignoring operations until after launch

Countermove: design monitoring, incident response, and model lifecycle management from the start.

Pitfall: Over-centralizing innovation or over-decentralizing standards

Countermove: standardize platforms and governance, allow teams to innovate within guardrails.

Pitfall: Building AI that lives outside workflows

Countermove: integrate AI into the process where decisions are made, then measure adoption and outcomes.

Signals you are on track

- AI initiatives reuse shared data products and platform capabilities.
- Stakeholders can trace AI outputs back to data sources and governance rules.
- Monitoring catches drift and quality issues before they become business incidents.
- Teams can deploy improvements safely without major outages or rewrites.
- Business users trust the system enough to change behavior, not just admire it.
- Scaling from one use case to the next gets faster, not slower.

> Design for change, not for perfection.

What I would do Monday morning

1. Select one high-value use case and define the target workflow and KPIs.
2. Identify the minimum viable data products required and assign ownership.
3. Define governance gates: privacy, risk, ethics, and approval decision rights.
4. Choose an architecture approach: Agile EA for iteration, TOGAF elements for rigor where needed (Jager, 2025).
5. Design integration into the workflow, not just a standalone model endpoint.
6. Establish monitoring requirements: drift, quality, performance, and incident response.
7. Confirm scalability patterns: cloud elasticity, microservices boundaries, and load handling (Enterprise IT Services, 2025; Jager, 2025).
8. Document the "reuse plan" so the next use case benefits from this investment.

Key takeaway

AI-driven enterprise architecture is how you turn AI from an experiment into an enterprise capability. Frameworks create discipline, scalable patterns create resilience, and workflow integration creates value.

When all three are designed intentionally, AI becomes a platform for continuous improvement, not a collection of disconnected initiatives.

Bridge to the next chapter

Now that we have the architectural foundation, the next challenge is execution at scale: how do you implement AI-driven enterprise architecture in a way that is secure, governed, and actually adopted?

In the next chapter, we will move from design principles into implementation strategies, focusing on sequencing, governance mechanisms, and operational practices that keep AI delivering value over time.

Microservices are a strategy only when ownership is clear.

CHAPTER 5

Managing Organizational Change for AI Adoption

The rollout that looked perfect on paper

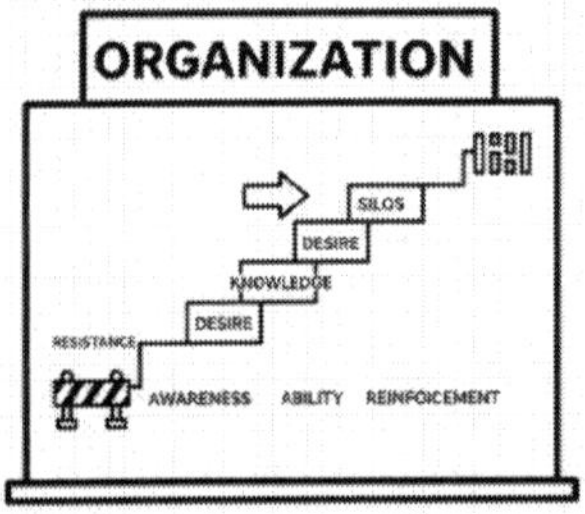

A leadership team invests in an AI tool that should reduce cycle time and improve decision quality.

The vendor demo is strong. The business case is approved. The implementation plan is detailed.

Three months later, adoption is inconsistent.

A few teams love the tool. Others avoid it quietly. Some employees complete training but continue using old spreadsheets.

Managers report mixed results and rising frustration. The tool itself works, yet the organization does not.

If this feels familiar, here is the uncomfortable truth: AI adoption is less about installing technology and more about changing behavior.

The human side of change is often the deciding factor between measurable impact and stalled projects.

This chapter shows you how to manage that change with intention. We will address cultural resistance, introduce proven change management approaches, and focus on employee engagement and measurement so AI becomes a trusted capability rather than a recurring struggle.

Understanding and overcoming cultural challenges in AI adoption

Resistance to change often starts as fear

When people hear "AI," many translate it into "automation," and then into "job loss." Even when leadership has no such intent, uncertainty fills the gap. Rumors travel faster than project plans.

Imagine a finance department where leaders introduce an AI-driven reporting tool with little context. Some employees assume their work is being replaced, while others disengage and avoid training because they do not see a future for themselves in the new workflow.

Productivity dips and managers interpret the learning curve as "people are resistant," when the real problem is unaddressed anxiety.

The countermove is direct, early communication that answers hard questions:

- What is changing and what is not?
- How will roles evolve?
- What skills will become more valuable?
- What support will be provided?
- How will success be measured fairly?

When employees see a plan for their growth, fear begins to shift into curiosity.

Lack of trust in technology is a rational response

Skepticism is not a personality flaw. It is often earned.

If teams do not understand how an AI tool works, or if early outputs look inconsistent, trust breaks quickly. A customer service team using an AI response assistant might see recommendations that feel off base.

If they cannot understand why, they will work around the system, override it manually, and tell each other it is unreliable.

Trust grows when leaders make the system understandable and safe to use:

- Explain what the AI can do and what it cannot do
- Show how decisions are made, at the right level of detail for the audience
- Run pilots that include frontline input, not just leadership approval
- Start with low risk use cases where errors are manageable
- Share early wins and lessons learned, including what did not work

A useful leadership question is: **Do my people trust this system enough to change how they work?** If the answer is no, adoption will stall no matter how strong the model is.

Build trust before you build dependence.

Silos create duplicated work and conflicting outcomes

Siloed organizations often launch AI initiatives in parallel without realizing it. Marketing and claims in an insurer may commission similar analytics work from different vendors, using different data definitions and reporting styles.

The result is duplicated cost, conflicting insights, and confusion about what is true.

The fix is not endless coordination meetings. The fix is structural:

- Shared data standards and governance
- Cross-functional prioritization of AI use cases
- A portfolio view of AI initiatives, not isolated projects
- Shared communities of practice to spread patterns and reuse

Leaders can ask: **Where are AI initiatives isolated, and what value would we unlock by connecting them?**

Entrenched processes and risk avoidance slow learning

Some organizations are so cautious that they eliminate the ability to experiment. Approval chains stretch for months, and by the time a pilot is approved, the opportunity has shifted.

A manufacturer considering AI-driven logistics optimization may lose momentum because every change must pass through lengthy reviews. Meanwhile, a competitor runs controlled experiments, learns quickly, and captures advantage.

The goal is not reckless innovation. The goal is **controlled learning**.

You can reduce friction by creating:

- A lightweight pathway for low-risk pilots
- Clear guardrails for privacy, security, and ethics
- Fast feedback loops that allow adaptation
- Visible recognition of responsible experimentation

Celebrating small wins signals that learning is valued, not punished.

Implementing change management strategies for AI initiatives

Good intentions do not change behavior. Structure does.

Change management models provide that structure. Two widely used approaches are Kotter's 8-Step Change Model and the Prosci ADKAR model. Both can be applied effectively in AI adoption when they are translated into practical actions.

Kotter's 8-Step Change Model in AI transformations

Kotter starts with urgency for a reason. If people do not feel why change matters now, they will default to familiar behaviors.

In an AI-driven forecasting initiative, leaders can establish urgency by connecting market disruption, customer expectations, and performance gaps to the need for better prediction and faster planning (Prosci, 2024).

From there, the model emphasizes:

- Building a coalition of credible champions
- Communicating a clear vision in plain language
- Removing obstacles, including workload and capability gaps
- Creating early wins and sharing them widely
- Embedding new habits into performance expectations and daily routines (Prosci, 2024)

Early wins matter because they convert abstract belief into lived proof. When one region processes orders 20 percent faster with AI support, it becomes easier for other teams to imagine success and commit to change.

The Prosci ADKAR model for employee-focused change

Where Kotter focuses on organizational movement, ADKAR focuses on the individual journey. ADKAR is helpful because adoption happens one person at a time.

ADKAR includes:

- **Awareness:** People understand why AI is being introduced and what problem it solves.
- **Desire:** People see how it benefits them and their team.
- **Knowledge:** People learn how to use the tool and the new workflow.
- **Ability:** People can apply the knowledge in real work with support.
- **Reinforcement:** Leaders reward and reinforce the new behavior so it sticks (Brusati, 2024).

In practice, desire often grows when employees understand that AI removes low-value tasks and creates space for higher-value work.

An HR team skeptical of AI screening tools may become more open when leaders show that the tool reduces repetitive work and frees time for interviewing, coaching, and candidate experience improvements (Brusati, 2024).

Organizational culture assessments as decision tools

Rolling out AI without assessing readiness is like building on sand.

Readiness surveys, interviews, and audits help you understand where trust is high, where change fatigue is present, and where previous experiences with technology shaped skepticism.

One practical outcome is smarter sequencing: start where readiness is higher, create visible wins, then expand.

Assessments also reveal strengths, such as frontline process insights or early champions who can support adoption.

Training and development to build confidence

Training must go beyond "click here, do this." AI changes how decisions are made and how work is performed, so training should cover both technical and behavioral components:

- How the tool works at a conceptual level
- What data it uses and what safeguards exist
- What the new workflow looks like end to end
- When to trust the system and when to override it
- How to escalate issues and report failures

Scenario-based training is especially effective. When teams role-play handling customer frustration during a chatbot error, they build practical confidence and reduce anxiety.

Soft skills training matters too: communication during change, comfort with ambiguity, and resilience under shifting priorities.

This combination, structured models, readiness assessments, targeted training, creates a durable foundation for adoption (Prosci, 2024; Brusati, 2024).

Engaging employees and evaluating change success in the AI journey

Once foundations are in place, engagement becomes the accelerant.

Involve employees early, not only at the end

People support what they help shape. Surveys and listening sessions create early insight into concerns and process realities.

Anonymous channels can surface questions that people hesitate to ask openly.

Cross-functional committees also help, especially when AI touches multiple workflows. When frontline employees participate in design choices, the solution becomes more usable and adoption increases.

Maintain transparency through ongoing dialogue

Town halls, Q&A sessions, and regular project updates reduce uncertainty.

A simple practice that works well is to share three things consistently:

- What is happening next
- What we learned in the last iteration
- What we are changing based on feedback

When employees see that their feedback changes the roadmap, trust increases. This responsiveness is also linked to psychological safety and healthier transitions during AI workplace changes (Soulami et al., 2024).

Recognize progress and make learning visible

Recognition builds momentum, especially during the messy middle of adoption.

Celebrating milestones, such as improved cycle time or increased usage, signals that the organization values progress and learning. Events such as internal demo days can turn isolated wins into shared confidence.

Use AI champions to accelerate adoption

Champions are employees who learn early, help peers, and represent frontline realities to leadership. They make change personal and practical.

Champions are especially effective when they come from multiple functions, HR, operations, customer service, and IT, because peer-to-peer support reduces fear and increases credibility (The Role of AI in HR Continues to Expand, 2025).

Build continuous feedback loops

Feedback should be continuous, not occasional:

- Pulse surveys on confidence and trust
- Focus groups on workflow friction
- Usage analytics to detect drop-offs
- Open forums for questions and concerns

When feedback highlights problems, respond visibly. For example, if adoption drops because users do not understand the tool's limits, short explainer videos and peer coaching can restore confidence.

Visible responsiveness strengthens trust and improves outcomes (Soulami et al., 2024).

Measure change success with a balanced scorecard

Adoption is measurable. Trust is measurable. Business impact is measurable.

A practical measurement set includes:

- Training completion rates
- Active usage rates and feature adoption
- Employee confidence and satisfaction signals
- Process performance improvements (cycle time, error reduction)
- Customer impact metrics where relevant
- Incident rates and escalation patterns

Tracking these measures creates early warning signals and helps leaders adjust before small issues become large failures (The Role of AI in HR Continues to Expand, 2025).

> Fear drops when people can see their role in the future.

Executive Lens: what leaders must get right

AI adoption stalls when leaders treat change management as optional.

As an executive sponsor, insist on:

- Clear messaging about role impact and skill development
- Budget allocated for change management, not only technology
- Named accountability for adoption outcomes
- Visible involvement from business leaders, not only IT
- A plan for trust, transparency, and escalation when AI fails
- Reinforcement mechanisms, performance metrics, recognition, and learning routines

The organization follows what leaders consistently reward.

Architecture and Delivery Lens: change management meets operating reality

Architects and delivery leaders can support adoption by designing systems that are usable, explainable, and integrated.

Key adoption-supporting design choices include:

- Transparent user experiences that show confidence levels and rationale where possible
- Clear fallbacks when AI is uncertain or fails
- Monitoring that detects drift and quality issues before users lose trust

- Integration into existing workflows so adoption does not require extra steps
- Support processes for feedback, incident handling, and continuous improvement

When the system is designed for trust and usability, change management becomes easier because the tool supports the human journey.

Common pitfalls and how to avoid them

Pitfall: Announcing AI without addressing job anxiety

Countermove: communicate role evolution, skill pathways, and support early.

Pitfall: Treating training as a one-time event

Countermove: provide ongoing coaching, champions, and scenario practice.

Pitfall: Ignoring skepticism as "negativity"

Countermove: treat skepticism as input, improve transparency, show early wins.

Pitfall: Pilots designed without frontline input

Countermove: include users in design and iterate based on lived workflow reality.

Pitfall: Measuring only usage or only ROI

Countermove: measure adoption, trust, workflow performance, and business outcomes together.

Signals you are on track

- Employees can explain why AI is being adopted and how it helps.
- Adoption rises steadily after launch rather than flattening.
- Champions are active and sought out by peers.
- Feedback loops lead to visible improvements.
- Business outcomes improve in parallel with usage, not years later.
- Trust increases over time, shown through reduced manual workarounds.

What I would do Monday morning

1. Identify the top two adoption risks for your next AI rollout and name owners.
2. Run a quick readiness assessment in impacted teams and adjust sequencing.
3. Create a clear message on role impact and skill development and deliver it live.
4. Establish a champions network across functions and train them early.
5. Build scenario-based training for real workflows, including failure handling.
6. Create an adoption dashboard that includes usage, confidence, and outcomes.
7. Schedule feedback checkpoints at weeks 2, 4, 8, and 12 post-launch.
8. Plan recognition moments tied to measurable progress, not just effort.

Key takeaway

AI adoption succeeds when leaders manage the human side with the same discipline they apply to technology. Address fear directly, build trust through transparency, break silos with shared ownership, and use structured change models to guide behavior change.

When people feel supported and the system is designed for usability, AI becomes a capability the organization embraces, not a tool it tolerates.

Bridge to the next chapter

Now that we have addressed how to lead people through AI adoption, the next question becomes governance: how do you set the policies, controls, and ethical guardrails that make AI innovation sustainable?

In the next chapter, we will focus on AI governance and responsible use, including risk management, compliance, and trust-building mechanisms that protect the organization while enabling progress.

Adoption is a leadership problem before it is a training problem.

CHAPTER 6

Ethics and Governance in AI Implementation

The moment a "successful" AI project becomes a leadership problem

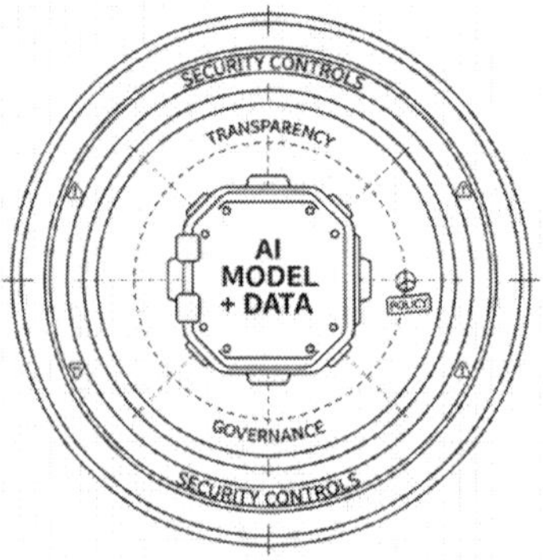

An AI capability goes live and early results look strong. Cycle time drops. Teams report faster decisions. Leadership feels momentum.

Then a customer complaint lands on the CEO's desk.

A customer believes they were treated unfairly by an automated decision. A regulator asks for documentation. An internal audit wants to know who approved the model, what data it used, and how bias was assessed. Security asks whether the model can be attacked or manipulated and HR asks whether employees understand how to use the tool responsibly.

In that moment, AI stops being a technology initiative and becomes a trust initiative.

Ethics and governance are how you protect trust while still moving fast enough to compete.

This chapter covers how to build ethical guidelines, secure AI systems against modern threats, and establish governance structures that keep AI compliant, accountable, and sustainable.

Why ethics and governance are now central to AI value

AI adoption is accelerating, and so are expectations around how organizations use it responsibly. Customers expect fairness and employees expect transparency. Regulators expect documentation and controls, and leaders need confidence that AI decisions align with company values and legal requirements.

The practical point is this: governance is not a brake on innovation. When done well, it is what allows you to scale safely.

Ethics and governance help you answer questions like:

- Are we using AI in ways that align with our values?
- Can we explain and defend AI-driven decisions to regulators and customers?
- Are we protecting privacy and sensitive data appropriately?
- Are we detecting and correcting bias before it becomes a headline?
- Are we resilient against attacks that target AI models and data?

Establishing ethical guidelines for AI use

Make ethics operational, not aspirational

An ethical framework is only useful if it changes day-to-day decisions. Many organizations formalize this through AI ethics review boards that include legal, technical, and policy perspectives, or by publishing AI principles that define expectations for fairness, transparency, and responsible data use (Mucci & Stryker, 2024; Papagiannidis et al., 2025).

These commitments matter because they create accountability. They also provide a reference point when teams face trade-offs under pressure.

A practical ethical framework typically covers:

- Responsibility and accountability for outcomes
- Transparency and explainability
- Fairness and bias mitigation
- Privacy and data protection
- Human oversight and escalation paths

Transparency builds trust and reduces risk

Transparency is one of the most effective governance tools because it makes AI visible and reviewable. When stakeholders can understand how AI is used, they are more likely to trust outcomes and less likely to assume the worst when something goes wrong (Mucci & Stryker, 2024).

Practical transparency mechanisms include:

- User-facing explanations of recommendations or decisions where appropriate
- Documentation such as model cards and decision logs
- Traceable records of data sources and data changes
- Feedback and complaint channels so people can question or contest outcomes

Transparency also supports audits and compliance reviews, especially in privacy contexts such as GDPR (Mucci & Stryker, 2024).

Fairness and bias mitigation are business issues, not just technical issues

Bias is not only an ethical risk. It is a reputational and regulatory risk. Biased outcomes can affect hiring, lending, personalization, and access to services.

When bias is left unaddressed, organizations can harm customers and employees while exposing themselves to penalties and public backlash (Mucci & Stryker, 2024; Papagiannidis et al., 2025).

Practical bias mitigation approaches include:

- Bias audits on training data and model outputs
- Fairness metrics applied during development to detect uneven outcomes
- Real-world pilot testing across different user groups
- Continuous monitoring after deployment, since bias can emerge over time

The leadership question to keep asking is simple:

Are outcomes fair across the groups we serve, and can we prove it?

> Fairness is not a checkbox.
> It's a continuous test.

Privacy and data protection are foundational

AI systems often depend on sensitive data. That means **privacy-by-design** is a baseline requirement, not a feature. Effective practices include consent and data minimization, strong access controls, and encryption of data in transit and at rest (Mucci & Stryker, 2024).

Regulations like GDPR raise expectations for how personal data is collected, used, and protected. Treating privacy as a core value supports compliance and strengthens customer trust (Mucci & Stryker, 2024).

Ethical guidelines create the foundation, but they are not enough on their own. AI systems face security threats that can undermine fairness, transparency, and trust. That brings us to the next layer.

Security challenges of AI systems

AI security requires more than traditional cybersecurity. Machine learning models introduce unique vulnerabilities that can be exploited directly.

Adversarial attacks and manipulation risks

AI systems can be targeted through adversarial attacks, where inputs are manipulated to cause harmful mistakes. A fraud detection system might be tricked by carefully crafted transactions. A hiring system might be influenced through poisoned training data that changes model behavior in subtle ways (Stepan Ilyin, 2025).

These risks are difficult because they can be hard to detect and evolve quickly as attackers adapt.

Continuous monitoring beats "set and forget"

Because threats change, AI security must be ongoing. Vulnerability assessments, penetration tests, and continuous monitoring help detect anomalies and attempted intrusions early. Static controls alone rarely keep pace (Stepan Ilyin, 2025).

Data security and access control are non-negotiable

AI models often require access to large datasets, sometimes containing highly sensitive information. Strong encryption, role-based access control, and audit trails reduce the risk of data exposure and support accountability (Jada & Mayayise, 2023).

Practical controls include:

- Encryption for stored and transmitted data
- Tight permissions for model parameters and datasets
- Logging and audit trails for data and model access
- Multi-factor authentication for sensitive systems

Human oversight remains a major control point

Many failures trace back to human behavior: unverified data used for training, poor access hygiene, weak monitoring, or unclear escalation paths. Training and simulations help teams understand security responsibilities and reduce accidental exposure (Stepan Ilyin, 2025).

Security-by-design must be part of architecture, not an add-on

Organizations that deploy AI without threat modelling often expose both data pipelines and model interfaces. A customer service chatbot, for example, can become an attack surface if APIs are not properly restricted.

Security-by-design includes adversarial testing, anomaly detection, and automated response mechanisms, and it requires security specialists to be involved early (Jada & Mayayise, 2023).

Security is closely tied to regulatory requirements. GDPR and HIPAA, for example, require traceable controls over personal data used or generated by AI systems. Detailed documentation, regular audits, and clear incident procedures support compliance and build stakeholder confidence (Stepan Ilyin, 2025).

Governance frameworks and compliance navigation for AI projects

Governance is how you turn ethics and security into repeatable practice. It clarifies ownership, decision rights, and accountability across the AI lifecycle.

Define roles and responsibilities clearly

AI governance breaks down quickly when ownership is unclear. When an AI system produces biased or harmful outcomes, leaders must know who is responsible for auditing, remediation, and decision-making (Mucci & Stryker, 2024).

Common governance roles include:

- AI project lead for delivery and ongoing monitoring
- Data stewards for data integrity, privacy, and lifecycle management
- Compliance liaisons for regulatory interpretation and documentation
- Executives for strategic alignment and accountability
- Security partners for threat modelling and controls

Clarity prevents fragmented responsibility and accelerates response when issues arise.

Use governance boards for oversight and "pause points"

Formal governance boards bring oversight across the AI lifecycle. Effective boards include technical experts, legal professionals, business leaders, and ethics perspectives. Their role is to review risk assessments, audit results, and incidents, and to enforce gates that prevent unsafe deployment.

Examples such as IBM's AI Ethics Board show how governance can provide structured review for transparency, fairness, and accountability, and prevent high-risk projects from progressing without remediation (Mucci & Stryker, 2024).

Governance boards are also critical during incidents. When an AI system behaves unexpectedly, the board can convene quickly, pause deployment if needed, and guide transparent investigation and communication.

Navigate a complex regulatory landscape proactively

Compliance is not a single checklist. Organizations must manage a patchwork of requirements such as GDPR, CCPA, and sector-specific obligations in banking, healthcare, and employment. Companies that track regulatory change proactively, through legal partnerships and internal policy updates, are better positioned to adapt without halting progress (GAN Integrity, 2024; Mucci & Stryker, 2024).

Practical approaches include:

- Regular internal policy reviews
- Mock audits and scenario training
- Documentation standards that support explainability and traceability
- Participation in industry forums to monitor emerging expectations, including work such as the OECD's efforts around explainable AI

When roles, boards, and regulatory awareness operate together, governance becomes a capability that supports speed and safety, not a bottleneck (GAN Integrity, 2024; Mucci & Stryker, 2024).

Executive Lens: what leaders must insist on

If you sponsor AI initiatives, these are the questions that protect the organization and enable scale:

- What values and ethical principles govern this use case?
- What is our transparency posture, and how do we handle challenges?
- How are fairness and bias assessed before and after deployment?
- What privacy obligations apply, and how are they implemented?

- What is the security threat model, and who owns mitigation?
- Who has decision rights to approve, pause, or retire the system?
- What documentation will we produce for audit and regulatory review?

Ethics and governance become real when executives require clear answers and fund the controls to deliver them.

Architecture and Delivery Lens: making governance executable

Architects and delivery leaders turn governance into working mechanisms. Key design choices include:

- Traceability from data sources to model outputs
- Monitoring for drift, performance issues, and anomalous behavior
- Access controls and audit trails across data and model components
- Clear workflows for incident response and escalation
- Documentation artifacts designed for review, not just storage

This is where responsible AI becomes operational, not theoretical.

Common pitfalls and how to avoid them

Pitfall: Publishing ethical principles but not changing delivery behavior

Countermove: embed ethics checkpoints and governance gates into the lifecycle (Mucci & Stryker, 2024).

Pitfall: Treating transparency as a marketing statement

Countermove: create real documentation, feedback channels, and explainability practices.

Pitfall: Bias checks that happen only before go-live

Countermove: monitor fairness and outcomes continuously (Papagiannidis et al., 2025).

Pitfall: Security applied late

Countermove: threat model early and build security-by-design into the architecture (Jada & Mayayise, 2023).

Pitfall: Unclear accountability during incidents

Countermove: define roles, decision rights, and escalation paths before deployment (GAN Integrity, 2024).

Signals you are on track

- Teams can explain AI decisions and provide documentation when asked.
- Bias and fairness are tested and monitored, not assumed.
- Privacy controls are embedded in workflows, not bolted on.
- Security monitoring detects anomalies and triggers response quickly.
- Governance boards meet regularly and are empowered to pause systems.
- Regulatory and audit requests can be handled without panic.

What I would do Monday morning

1. Create or refresh a short set of AI principles that match company values.
2. Establish an AI governance structure with named roles and decision rights.
3. Define minimum documentation standards for every AI use case.
4. Implement a bias and fairness assessment process for development and production.
5. Require privacy-by-design controls and confirm alignment with GDPR and relevant regulations.
6. Build a threat model for your highest-risk AI systems and prioritize mitigations (Stepan Ilyin, 2025).
7. Set up monitoring for drift, anomalies, and incident reporting workflows.
8. Schedule mock audits and scenario exercises to test readiness (GAN Integrity, 2024).

Key takeaway

Ethics and governance are not separate from AI delivery. They are what make AI scalable and sustainable. When you embed transparency, fairness, privacy, and security into architecture and operating mechanisms, you build trust with customers, employees, and regulators while still moving quickly enough to compete (Mucci & Stryker, 2024; Papagiannidis et al., 2025).

Bridge to the next chapter

With ethical guidelines and governance in place, the next challenge is proving sustained value over time.

In the next chapter, we will focus on measurement and optimization: how to track outcomes, refine AI performance, and build continuous improvement loops that keep AI-driven transformation delivering results long after the initial launch.

> If you can't explain it, you can't defend it.

CHAPTER 7

Measuring AI Impact on Business Outcomes

The board question that every AI leader eventually gets

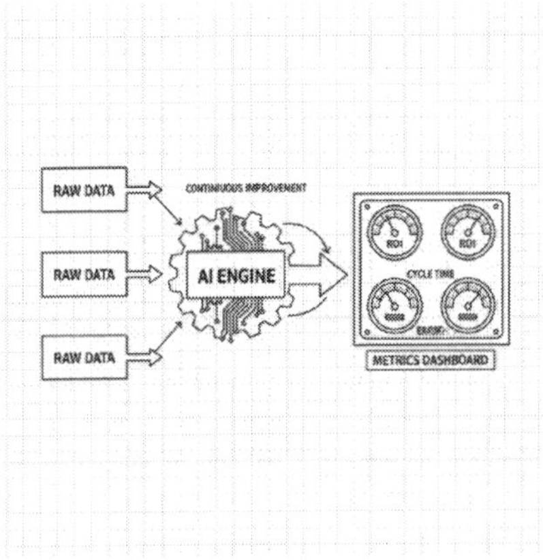

The AI program lead finishes an update with a confident summary: model performance is improving, adoption is growing, and teams are excited.

Then a board member asks a simple question: "So what changed in the business?"

Not "How accurate is the model?"
Not "How many users logged in?"
Not "How many use cases are in the pipeline?"

What changed in the business.

That question is where many AI initiatives stumble, not because they lack value, but because they lack measurement discipline.

AI impact is often real but hard to prove unless you define success in business terms, track it consistently, and communicate it clearly.

This chapter shows you how to do that. We will define practical KPIs, translate them into ROI, and build a measurement and communication approach that keeps AI investment credible and scalable.

Measuring AI success means measuring more than the model

AI introduces new capabilities: automation, prediction, personalization, and decision support. Traditional business metrics like profit and market share still matter, but they move slowly and are influenced by many factors. AI measurement works best when you connect three layers:

- **Business outcomes**: revenue, cost, risk, customer experience, employee productivity
- **Operational indicators**: cycle time, error rate, throughput, compliance outcomes, adoption
- **Technical performance**: accuracy, false positives, latency, drift

If you measure only technical performance, you will struggle to prove value. If you measure only business outcomes, you may miss early warning signs. You need all three, with business outcomes leading the story (Understanding AI Strategy: Key Principles for Success in Today's Digital Era, 2024).

Defining KPIs for AI success

Start with KPIs that match the promised benefit

KPIs should reflect the specific change AI is supposed to create: faster processes, higher quality decisions, lower cost, improved experiences, or reduced risk.

Generic metrics rarely capture the real impact of an AI initiative, especially early on (Understanding AI Strategy: Key Principles for Success in Today's Digital Era, 2024).

Example: customer service chatbot

Revenue may not change immediately. Better KPIs include:

- Average response time
- First-contact resolution rate
- Escalation rate
- Customer satisfaction after interaction
- Agent productivity in complex cases (Jacob, 2025)

The practical test is this: if the AI disappeared tomorrow, would these KPIs worsen in a measurable way? If yes, you are measuring the right things.

Ensure every KPI is specific, measurable, and relevant

Strong KPIs usually share three traits:

- **Specific** to the intended AI benefit
- **Measurable** with reliable data and consistent definitions
- **Relevant** to the project and to how the organization operates (Jacob, 2025)

This matters because AI initiatives often fail to gain executive support when KPIs are vague, inconsistent across teams, or impossible to measure cleanly.

Tie project KPIs to enterprise priorities

AI initiatives stay funded and supported when leaders can see a direct link to strategic goals such as growth, resilience, operational excellence, or risk reduction.

A supply chain forecasting model, for example, should connect to KPIs such as:

- Inventory turnover
- Cost per shipment
- Order fulfillment accuracy
- Stock-out rate
- Expedited shipping costs

When the KPI story clearly supports enterprise goals, stakeholders focus on outcomes, not hype (Understanding AI Strategy: Key Principles for Success in Today's Digital Era, 2024).

Combine quantitative and qualitative measures

Not all impact appears in spreadsheets. AI can change how work feels, how confident employees are, and how much trust customers place in decisions.

Consider automated route planning in delivery operations. Efficiency metrics may improve, while drivers report frustration with unpredictable routes. If you track only speed and cost, you might miss the adoption and trust risk that will eventually reduce value.

> A model metric is not a business metric.
> Connect the chain.

Add qualitative signals such as:

- User satisfaction surveys
- Feedback from frontline teams
- Customer sentiment and complaints
- Confidence ratings for AI recommendations (Jacob, 2025)

These inputs help you correct early and protect long-term adoption.

Use benchmarks to define what "good" looks like

Benchmarks anchor your measurement. They can be:

- **Internal**: pre-AI baselines such as cycle time, error rate, or cost
- **Pilot-based**: performance during limited rollout
- **External**: industry averages or peer comparisons (Jacob, 2025)

Benchmarks also prevent what many teams fall into: success targets that become outdated while the market shifts.

Scheduled KPI reviews help ensure the organization is not optimizing for yesterday's definition of value (Understanding AI Strategy: Key Principles for Success in Today's Digital Era, 2024).

Example: A tech firm initially optimized for faster response time in support. Later, customer feedback made it clear that thorough, accurate answers mattered more than speed.

The KPIs shifted accordingly, protecting retention and satisfaction rather than chasing a metric that no longer matched customer expectations (Understanding AI Strategy: Key Principles for Success in Today's Digital Era, 2024).

A practical KPI set you can reuse

Here is a balanced set many enterprises use across AI initiatives:

Business outcomes

- Revenue lift, conversion, retention
- Cost reduction, productivity gains
- Risk reduction and compliance outcomes

Operational outcomes

- Cycle time, throughput, backlog trends
- Error rates, rework rates
- Adoption, usage frequency, workflow completion

Technical outcomes

- Precision and false positives
- Latency and uptime
- Drift and retraining frequency

When KPIs are well defined and kept current, measurement becomes a leadership tool, not just reporting.

Calculating ROI from AI investments

KPIs show progress. ROI proves value in terms executives and finance teams can act on.

ROI for AI initiatives generally follows a familiar formula:

ROI = [(Total Benefits – Total Costs) / Total Costs] x 100

The challenge is not the equation. The challenge is disciplined definition of costs and benefits.

Step 1: capture full cost, not just the build

AI costs include more than tools and software. A complete view often includes:

- Data work: quality, governance, access, labelling
- Integration and platform changes
- Security and privacy controls
- Training and change management
- Ongoing monitoring, maintenance, and model updates

Step 2: quantify benefits across cost, speed, quality, revenue, and risk

Measurable benefits often appear as:

- Reduced labor and manual effort
- Lower error rates and rework costs
- Faster process completion and higher throughput
- Reduced downtime or operational losses
- Revenue lift through personalization or improved conversion

One common approach is to compare pre-AI and post-AI performance and translate the delta into financial terms.

Example: If automation reduces data entry errors by 80 percent and saves $100,000 per year in rework, that becomes a clear ROI line item.

Step 3: separate "proof" from "potential"

Some benefits are direct and measurable. Others are strategic and should be framed carefully.

A practical way to keep credibility:

- Treat measurable cost and revenue changes as the core ROI proof
- Treat strategic benefits, such as agility and improved decision quality, as additional value supported by qualitative evidence and trend data

Case studies and measurement guides emphasize the value of tracking both types while keeping the financial story grounded in evidence (Babina et al., 2024; 2025 Volume 5 How to Measure and Prove the Value of Your AI Investments, 2025).

Step 4: use testing to isolate AI's contribution

When revenue impact is involved, A/B testing or control groups can help isolate the effect of AI.

An ecommerce firm can test an AI recommendation engine by showing half of shoppers a standard experience and half the AI-driven flow, then measuring differences in spend, conversion, and retention.

This improves credibility because it separates AI impact from seasonal shifts, pricing changes, or marketing campaigns.

Step 5: create a repeatable ROI checklist

Use this checklist to make ROI estimation consistent:

- List all project-related costs, including ongoing costs
- Gather baseline performance data
- Define benefit metrics linked to KPIs
- Collect post-launch data at planned intervals
- Convert changes into financial terms
- Document assumptions and ranges to keep the model honest

Effective communication and continuous improvement

Measurement creates value only when it creates decisions. Communication turns measurement into momentum.

Tell the story behind the data

Numbers are necessary, but they are rarely memorable on their own. Combine metrics with a short, concrete story about what changed in work and outcomes.

A time-saving story might sound like: "With AI-supported onboarding, new hires begin contributing sooner because they spend fewer days navigating internal systems."

The point is to connect the metric to **real business experience**.

Tailor the message to the audience

Different stakeholders care about different parts of the value story.

- **C-suite and boards:** business outcomes, risk, competitive advantage, governance posture
- **Finance:** cost categories, assumptions, payback, sensitivity
- **Business unit leaders:** operational impact, adoption, workflow performance
- **IT and architects:** scalability, reliability, monitoring, integration learnings
- **Frontline teams:** how the tool helps them, what changes in daily work, where to get support

Technical leaders may want details on models and systems. Business leaders want the "so what."

Your job is to connect them without burying the point.

Use visuals that make impact obvious

Choose visuals that match the message:

- Bar charts for before and after comparisons
- Line charts for improvements over time
- Simple dashboards for ongoing KPI tracking
- Side-by-side comparisons for control versus AI-assisted groups

The goal is instant comprehension. If a leader needs five minutes to interpret the chart, the message is too complex.

Create two-way communication loops

Communication should invite feedback, not only deliver updates. Town halls, Q&A forums, and targeted feedback sessions help surface confusion or resistance early. Some organizations use sentiment analysis on surveys to identify patterns in concerns and adjust messaging and training accordingly (Swift, 2025).

Feedback loops also improve the AI systems themselves by highlighting where outputs are unclear, workflows are misaligned, or adoption barriers exist (Rachinger et al., 2019).

Continuous improvement is the point, not a bonus

AI performance and business needs change over time. Scheduled KPI reviews prevent the organization from optimizing for outdated goals. Continuous evaluation helps refine both the model and the operating process, ensuring AI stays aligned to business outcomes (Understanding AI Strategy: Key Principles for Success in Today's Digital Era, 2024).

Executive Lens: the questions that prove value

If you lead AI investment decisions, insist on answers to:

- What business outcome is this initiative accountable for?
- What are the primary KPIs, and what are the baselines?
- What adoption level is required for value to appear?
- What is the measurement plan for 3, 6, 12, and 24 months?
- What financial value is proven versus assumed?
- What will we change if the KPIs stall or adoption declines?

When leaders ask these questions consistently, AI measurement becomes part of how the organization runs, not an occasional report.

Architecture and Delivery Lens: measuring what makes AI sustainable

Architects and delivery leaders should ensure measurement includes:

- Model monitoring: drift, accuracy, latency, incidents
- Data quality measures and lineage tracking
- Workflow integration signals and usage patterns
- Reliability, scalability, and operational cost trends

These signals protect long-term value because they reveal early when the AI system is degrading or when the operating model is not keeping up.

Common pitfalls and how to avoid them

Pitfall: Measuring only model accuracy

Countermove: measure workflow and business outcomes first, then technical performance (Jacob, 2025).

Pitfall: Using KPIs that do not match the promised benefit
Countermove: choose KPIs tied directly to the intended operational and business change.

Pitfall: Treating adoption as a side metric

Countermove: track adoption as a primary value driver, not an afterthought.

Pitfall: ROI models that ignore ongoing costs

Countermove: include maintenance, monitoring, security, and retraining in total cost.

Pitfall: Reporting results without context

Countermove: pair metrics with stories, visuals, and clear assumptions.

Signals you are on track

- Leaders can explain business value without referencing technical details.
- KPIs show improvement in outcomes and adoption, not only usage.
- ROI is supported by baseline comparisons and documented assumptions.
- Measurement drives decisions, not just reporting.
- KPIs evolve as customer expectations and business priorities shift.

What I would do Monday morning

1. Define the top three business outcomes for AI this year and the KPIs that prove them.
2. Establish baseline metrics and confirm shared definitions across teams.
3. Build a balanced KPI dashboard: business, operational, and technical measures.
4. Choose one initiative and run a control group or A/B test where feasible.
5. Create a standard ROI worksheet that includes full costs and benefit categories.
6. Schedule reviews at 3, 6, 12, and 24 months with clear accountability.
7. Build a communication pack: one page for executives, one page for delivery teams, one page for frontline users.
8. Create a feedback channel and commit to visible responses based on what you learn.

Key takeaway

You prove AI value by measuring business outcomes, not by celebrating technical performance. Strong KPIs connect AI to operational change, ROI translates that change into financial terms, and clear communication turns measurement into sustained momentum.

Done well, measurement becomes the engine that keeps AI investment credible, scalable, and aligned to what matters most.

Bridge to the next chapter

Now that we can measure and prove AI impact, the next challenge is sustaining it over time.

In the next chapter, we will focus on continuous improvement and future readiness: how to keep architectures, teams, and governance evolving as AI capabilities and market demands change.

Make value visible, or value will be questioned.

CHAPTER 8

Future Trends in AI and Enterprise Architecture

The strategy offsite where the timeline suddenly shrinks

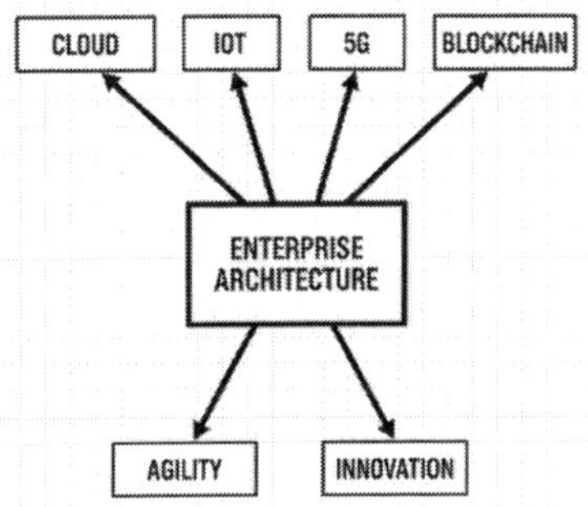

A leadership team sits in a strategy offsite looking at the next three years. The plan is sensible: modernize key platforms, improve data quality, and pilot a few AI use cases.

Then someone asks a simple question: "What if the competitive baseline changes faster than our plan?"

That is the real tension of the next era. Many forecasts suggest that, over the next five years, more than 80% of enterprises will increase AI investment as a core part of business strategy. The consequence is not just more technology spend. It is a new operating tempo.

Decisions that once waited for quarterly reports now move on real-time signals. Architecture that once prioritized stability alone now must balance stability with learning, adaptability, and rapid iteration.

This chapter is your guide to what is coming next and how to prepare. We will look at emerging technologies shaping AI, the expanding role AI plays in digital transformation, and how leaders can build a long-term vision that turns disruption into advantage.

The future is not one technology. It is a stack of technologies working together.

When leaders think about "future AI," they often focus on models. But in the enterprise, models only create value when the surrounding system makes them usable, secure, scalable, and integrated into workflows.

The trend to watch is convergence: cloud, IoT, edge computing, faster networks, and trust layers such as blockchain are combining into architectures where AI can learn continuously and act faster than traditional organizations can respond.

Emerging technologies shaping AI's future

Cloud and cloud-native AI as the default foundation

Cloud computing has moved from an infrastructure decision to a speed decision. Instead of waiting through long procurement cycles, organizations can scale compute and storage on demand, run experiments quickly, and deploy AI capabilities across regions with far less friction (Erickson, 2024).

This matters because AI development is iterative. Teams need environments where they can test, learn, adjust, and ship improvements without rebuilding everything each time.

What it looks like in practice: A retail chain can roll out an AI recommendation engine broadly and update models centrally while product teams collaborate across geographies in shared environments (Erickson, 2024). The architecture benefit is not only elasticity. It is repeatability and faster learning cycles.

Cloud-native approaches also reduce friction between business units and IT by enabling proofs of concept in shared environments for chatbots, document automation, and analytics pilots (Erickson, 2024). When the environment is ready, alignment and delivery speed improve.

IoT and edge data as the fuel for real-time learning

The Internet of Things is changing enterprise architecture by producing high-frequency operational data from sensors in equipment, logistics networks, and workplaces. This data is uniquely valuable because it reflects what is happening now, not what happened last month (Auday Al-Dulaimy et al., 2024).

Why this matters: AI becomes more responsive when it learns from current signals. Predictive maintenance is the classic example. Sensors monitoring vibration and energy usage allow AI systems to anticipate failures before downtime occurs. In logistics, connected trackers provide up-to-the-minute visibility across routes and handoffs (Auday Al-Dulaimy et al., 2024).

The architectural shift is that "edge to cloud" becomes a core design pattern, with governance and security controls spanning both.

5G as a catalyst for low-latency decisions

5G amplifies the impact of IoT and edge computing by reducing latency and making fast, distributed decisions more practical. When data moves in milliseconds, AI can act in near real time across mobile devices and remote environments.

This expands use cases where speed is essential: wearables triggering clinical alerts, vehicles reacting to road conditions, retailers delivering real-time in-store offers, and field engineers receiving predictive guidance while on-site.

The enterprise architecture implication is that decision-making becomes distributed. More intelligence moves closer to where the work happens.

Blockchain and trust layers for traceability and shared data ecosystems

Blockchain is increasingly discussed as a trust layer where tamper-evident records can strengthen traceability of data and decisions, especially in sensitive contexts such as financial services and supply chains.

When multiple organizations collaborate on shared AI ventures, a trust mechanism that logs data changes and model updates can reduce disputes and improve auditability. In sectors where provenance matters, this can support stronger governance and confidence in the integrity of training data and decision trails.

The practical takeaway: as AI systems influence higher-stakes outcomes, architectures will need stronger mechanisms for traceability, provenance, and audit readiness.

AI's expanding role in digital transformation

Emerging technologies create the foundation, but AI is reshaping how organizations run. The pattern is consistent: AI moves from isolated automation to enterprise-wide capabilities that change operating models.

Operational efficiency and resilience

AI-driven automation and predictive analytics continue to reduce manual effort, improve speed, and reduce downtime.

Research cited in Ghosh (2025) highlights predictive maintenance impacts such as substantial reductions in maintenance costs and improved reliability, reflecting earlier work such as Shrestha et al. (2019) (Ghosh, 2025).

Other research cited also points to meaningful operational cost reductions in organizations using AI effectively (Ghosh, 2025).

Supply chains illustrate the broader shift. AI-assisted routing and forecasting can improve logistics efficiency and reduce stockouts by shifting planning from reactive to proactive (Ghosh, 2025).

The enterprise architecture trend here is continuous feedback loops: operational data feeds models, models influence decisions, and outcomes feed back into learning.

Customer-centric innovation at scale

AI is increasingly the engine behind personalization and faster customer resolution. AI-powered chatbots and virtual assistants can reduce resolution times, support customer loyalty and lowering service friction (Ghosh, 2025).

Organizations are also learning to sense customer needs earlier by analyzing feedback, browsing behavior, and product usage patterns. This changes product development cycles, marketing timing, and service design.

Better, faster decision-making

The move from static reporting to real-time decision support is accelerating. Dashboards update continuously, forecasts adjust rapidly, and teams reallocate resources based on live signals.

Research cited in Ghosh (2025) suggests decision errors can decline meaningfully when AI-based decision support is used well, reinforcing the advantage of faster learning loops (Ghosh, 2025).

AI-first operating habits

The deeper shift is cultural. Organizations that embed AI into daily workflows begin to expect change. Teams get comfortable testing, learning, and iterating. Leaders rely more on evidence than opinion. Collaboration improves because shared data and shared metrics become the reference point (Kassa & Worku, 2025; Ghosh, 2025).

Preparing for disruption and building a long-term AI vision

Future readiness is less about predicting exactly what will happen and more about building an organization that can adapt quickly when reality changes.

Use AI to look outward, not only inward

Many companies first use AI to optimize internal operations. The next leap is using AI to detect external shifts early: market demand changes, supply chain disruption, emerging competitor moves, and changes in customer sentiment.

A retail organization can scan transactions and traffic patterns continuously, spotting early signals of demand shifts and adjusting inventory and marketing before losses accumulate (Marsh, 2024). In manufacturing, AI can help detect early indicators of supply disruption and enable faster rerouting or renegotiation (Marsh, 2024).

The architecture requirement is an external data strategy: what sources you monitor, how you validate them, and how you turn signals into decisions.

Competitive mapping becomes continuous

AI-powered benchmarking can track competitor pricing, feature changes, sentiment, and customer signals more continuously than manual research can. A financial services company can compare app features and customer sentiment against fintech competitors and surface emerging threats quickly for strategy review (Adamec, 2024).

This does not replace leadership judgment. It improves timing, which is often the difference between proactive moves and defensive reactions.

Treat the AI strategy as a living blueprint

Long-term AI success comes from a roadmap that evolves based on learning. That roadmap should include:

- Clear objectives tied to business results
- A portfolio view of AI initiatives, not scattered experiments
- A cross-functional steering mechanism that adjusts priorities
- Visible benchmarks for value and adoption
- Ongoing investment in data, governance, and skills (Marsh, 2024; Adamec, 2024)

This is how organizations avoid "pilot fatigue" and build a repeatable transformation engine.

Executive Lens: what leaders should focus on next

If you lead transformation, the future trends can feel endless. Here is what to anchor on:

- **Speed with control:** how to move fast without breaking trust
- **Architecture as leverage:** reusable platforms, data products, and governance that make the second use case easier than the first
- **Talent and operating model:** teams that can iterate continuously, not deliver once
- **Value proof:** measurement discipline that keeps investment credible
- **External sensing:** using AI to detect market shifts early, not just optimize internal processes

Your competitive advantage will increasingly be your ability to learn and adapt faster than your competitors.

Architecture and Delivery Lens: what to build for future readiness

For enterprise architects and delivery leaders, future readiness usually comes down to a few concrete capabilities:

- Cloud-native foundations that support rapid experimentation and scaling (Erickson, 2024)
- Edge and IoT integration patterns that handle high-frequency data safely (Auday Al-Dulaimy et al., 2024)
- Strong data governance and traceability as AI influences higher-stakes decisions

- Monitoring and feedback loops that keep models reliable as conditions change
- Modular architectures that allow safe iteration and replacement of components over time

The goal is not to chase every trend. The goal is to design an architecture that can absorb change without constant reinvention.

Common pitfalls to avoid

Pitfall: Treating future trends as a shopping list

Countermove: choose trends that directly support your strategy and operating model.

Pitfall: Scaling AI models without scaling governance and trust mechanisms

Countermove: design traceability, accountability, and monitoring as core architecture elements.

Pitfall: Over-centralizing innovation or letting every team reinvent the wheel

Countermove: standardize platforms and guardrails, enable teams to innovate within them.

Pitfall: Measuring activity instead of outcomes

Countermove: keep KPIs tied to business value and adoption, and review them regularly.

Signals you are prepared for what is next

- You can deploy and improve AI capabilities without major rebuilds.
- Data products have clear ownership and are reusable across use cases.
- Governance is strong enough to enable speed, not block it.
- Edge and cloud architectures support real-time decision needs where required.
- Teams and leaders are comfortable with continuous learning and iteration.
- You have visibility into external market signals, not only internal performance.

What I would do Monday morning

1. Identify the two to three future trends most relevant to your strategy: cloud scale, edge and IoT, low-latency decisions, or trust and traceability.
2. Map those trends to specific business outcomes and the workflows they will affect.
3. Assess your architecture readiness: data foundation, governance, integration patterns, and monitoring.
4. Create a one-year roadmap that prioritizes reusable capabilities over isolated use cases.
5. Establish an “external sensing” pilot: pick one market signal stream and build a lightweight dashboard for decision review (Marsh, 2024; Adamec, 2024).
6. Build a talent plan: upskilling, communities of practice, and operating model updates that support continuous delivery.

7. Schedule quarterly strategy and architecture refresh cycles to keep the roadmap current.

Key takeaway

Future AI advantage will come from convergence: cloud-native speed, IoT-driven real-time data, faster networks enabling edge decisions, and stronger trust mechanisms for traceability.

Leaders who treat AI as a continuous capability, supported by enterprise architecture and governance, will be positioned to adapt faster, innovate more confidently, and turn disruption into opportunity (Erickson, 2024; Auday Al-Dulaimy et al., 2024; Marsh, 2024; Adamec, 2024).

Bridge to the next chapter

Now that we have a view of where AI and enterprise architecture are heading, the next question is practical: how do you build an organization that keeps learning, improving, and sustaining transformation over time?

In the next chapter, we will focus on making AI-driven transformation durable through continuous improvement, talent development, and operating rhythms that keep progress alive long after the initial wave of implementations.

The future is composable.
Your architecture should be too.

CHAPTER 9

Case Studies of AI-Driven Transformation

"AI will save us from ourselves," the CEO said

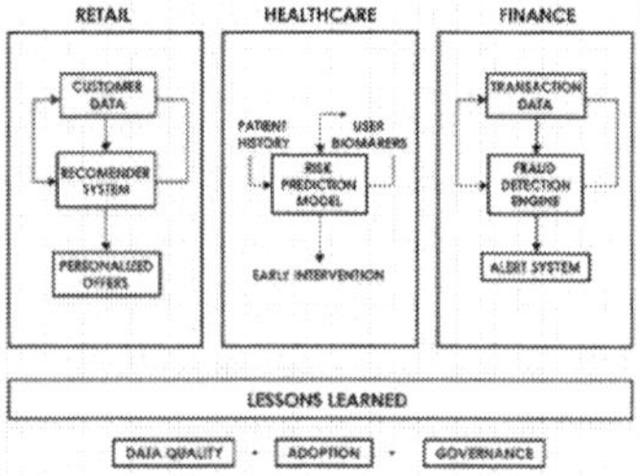

"AI will save us from ourselves, or so they say."

A CEO of a global retailer said that to me in a board meeting, frustration obvious. She was not doubting the technology. She was tired of seeing big promises collide with messy realities: unclear ownership, inconsistent data, low adoption, and pilots that never became enterprise capabilities.

If you have led transformation, you recognize the pattern. AI can absolutely drive measurable impact, but it is not magic.

It works when strategy is clear, data is trustworthy, teams collaborate across boundaries, and governance keeps decisions responsible and explainable.

This chapter brings those ideas to life through real-world examples. We will look at where AI delivered meaningful results, where it failed, and what leaders can learn to increase the odds of success.

What these case studies are really about

At first glance, these examples look like technology stories: recommendation engines, diagnostic models, robotics, fraud detection.

But the underlying success factors are almost always the same:

- A business problem defined in measurable terms
- Data quality and governance treated as foundational
- Workflow integration that changes how work is actually done
- Change management that builds trust and adoption
- Ongoing monitoring, improvement loops, and clear accountability

When these elements show up together, AI becomes a transformation lever. When they are missing, even strong models struggle to deliver.

Overview of successful AI deployments and common pitfalls

Retail: recommendations and forecasting that moved real metrics

Retailers have used machine learning to personalize shopping experiences by analyzing browsing history, purchases, and other signals to recommend products in real time. When well executed, this can lift basket size and improve repeat purchase behavior.

Recommendation systems also become stronger when marketing, product, and data science teams collaborate to refine data feeds and retrain models regularly, keeping relevance high as customer behavior shifts (AI Case Studies in Healthcare, Finance, Retail & More, 2020; Rashid & Kausik, 2024).

Retailers have also improved inventory forecasting by predicting local demand changes and generating low-stock warnings that help reduce stockouts and excess inventory.

The most important lesson is that these results tend to depend on clean data, integrated workflows, and rapid iteration, not just the model itself (AI Case Studies in Healthcare, Finance, Retail & More, 2020; Rashid & Kausik, 2024).

Common pitfall in retail: teams focus on model performance but fail to connect outputs to decisions. If store managers do not trust recommendations, or if replenishment workflows cannot act on predictions, value stalls.

Healthcare: decision support that improved speed and coordination

Healthcare organizations have used deep learning models to support radiology teams by flagging anomalies in scans, helping clinicians prioritize review and reduce missed findings.

Impact improves when AI outputs are integrated into clinical workflows, such as side-by-side review tools, rather than delivered as separate dashboards that clinicians do not use (Rashid & Kausik, 2024; AI Case Studies in Healthcare, Finance, Retail & More, 2020).

Predictive analytics has also been applied to readmission risk, supporting targeted follow-up by social workers and discharge planners.

Natural language processing-based assistants have been used for patient education and basic triage, reducing strain on staff and improving patient engagement signals such as no-show rates (Rashid & Kausik, 2024; AI Case Studies in Healthcare, Finance, Retail & More, 2020).

Common pitfall in healthcare: organizations underestimate the importance of trust and human override pathways. If clinicians cannot understand when to rely on AI and when to challenge it, adoption declines and patient safety concerns rise.

Manufacturing: computer vision and predictive maintenance that reduced waste and downtime

Manufacturers have used computer vision for defect detection on production lines, improving quality by identifying issues beyond human visibility and feeding fault patterns back into process improvement.

Predictive maintenance models using sensor data have also reduced unplanned downtime when maintenance shifts from calendar-based schedules to need-based interventions (Rashid & Kausik, 2024).

Successful programs often include workforce upskilling, cross-functional teams spanning IT and operations, and dashboards that translate AI outputs into operational actions.

Common pitfall in manufacturing: scaling is blocked when sensor data is inconsistent across plants, or when maintenance teams lack training and confidence to act on predictions.

The connecting thread across these successes

In retail, healthcare, and manufacturing, success depended less on “using AI” and more on building the conditions for AI to matter: trustworthy data, cross-functional coordination, workflow integration, training, and ongoing improvement loops (AI Case Studies in Healthcare, Finance, Retail & More, 2020; Rashid & Kausik, 2024).

Those same themes appear even more sharply in financial services, where regulatory and trust expectations are unforgiving.

Sector-specific insights: wins and failures in financial services

Financial services have been an early adopter of AI, driven by fraud risk, customer experience expectations, and competitive pressure from fintech entrants. Large firms have invested heavily in AI platforms to sharpen competitiveness (Vuković et al., 2025).

Fraud detection: from static rules to learning systems

Banks deploying AI-driven real-time transaction monitoring have reported significant reductions in fraud losses compared to traditional rule-based systems.

The advantage comes from continuously learning spending patterns, detecting subtle anomalies, and adapting as fraud tactics evolve (DigitalDefynd, 2024).

What leaders should notice: fraud detection success is rarely only an algorithm story. It is a data pipeline story, a monitoring story, and an operating model story. If you cannot retrain, monitor drift, and update controls quickly, the system degrades.

Lending and underwriting: faster decisions with explainability built in

Some financial organizations have applied AI to loan processing by combining structured and unstructured data using machine learning and NLP. Reported outcomes include reduced processing time and improved detection of high-risk applications while maintaining compliance through explainable outputs and human oversight (DigitalDefynd, 2024; Vuković et al., 2025).

What leaders should notice: speed alone is not the win. The win is faster decisions that remain defensible. Explainability and oversight are what keep the gains sustainable in regulated environments (Černevičienė & Kabašinskas, 2024).

Investment personalization and market intelligence

Investment organizations have used AI platforms combining machine learning with sentiment analysis and diverse datasets to improve portfolio insights and personalization under volatile market conditions.

Reported outcomes include improved performance in certain implementations when systems integrate multiple data sources and maintain compliance-friendly transparency (DigitalDefynd, 2024).

What leaders should notice: value comes from integrating diverse signals into decision processes, not just adding analytics. These systems succeed when they fit how investment decisions are governed and executed.

Where financial services AI projects fail, and why

Failure mode 1: launching customer-facing AI without readiness and iteration

Some institutions pushed chatbots and digital advisors at scale without testing adoption and customer expectations through incremental rollout. Customers experienced the tools as impersonal or confusing, uptake stalled, and the organization lost credibility.

The root cause is usually not the model. It is a failure to iterate with real users and refine workflows.

Failure mode 2: insufficient oversight and explainability

Firms have faced regulatory and reputational risk when automated decisions lacked transparency or adequate human oversight, including issues tied to biased outcomes in sensitive areas such as credit decisions (Vuković et al., 2025).

This is why human-in-the-loop design and explainability are not optional. They are core controls for responsible deployment (Černevičienė & Kabašinskas, 2024).

Failure mode 3: limited diversity of input and blind spots in risk assessment

Some high-profile missteps occur when teams do not include diverse perspectives during design and testing, leading to bias, unfair outcomes, and reputational backlash. These failures often reveal a governance gap more than a technical gap.

Action list for business leaders

If you want AI transformation that survives contact with reality, these actions consistently show up in successful organizations:

- Invest in data quality and governance from day one.
- Prioritize explainability and human oversight in automated decision-making, especially customer-facing processes.
- Roll out AI capabilities incrementally, collect feedback, then expand.
- Align business objectives with realistic expectations for short-term and long-term impact.
- Maintain active dialogue with regulators where required, and address ethical concerns proactively (Vuković et al., 2025; DigitalDefynd, 2024).

Benchmarking success and innovative use cases

Strong organizations do not measure AI success once. They benchmark continuously.

Benchmarking helps leaders evaluate whether AI investments improve operational performance, customer outcomes, and financial returns compared with internal baselines and industry norms (Dreyfus, 2025).

What to benchmark in practice

For customer support AI, common benchmarks include:

- Containment and deflection rates
- Resolution outcomes
- Agent-assist utilization

- Customer satisfaction and abandonment rates (Dreyfus, 2025)

Some industries report containment rates in the 60% to 80% range for common queries, which can provide an external reference point for comparable deployments (Dreyfus, 2025).

For broader programs, benchmarking also includes:

- ROI and cost-to-serve trends
- Quality and error rates
- Adoption by employees and customers

Compliance outcomes and audit readiness (Anchia, 2024; Dreyfus, 2025)

Why transparency and documentation matter for benchmarking

Benchmarking works best when organizations maintain records of model changes, data sources, algorithm adjustments, and user feedback. This supports compliance and makes performance shifts explainable, which increases stakeholder confidence (Dreyfus, 2025).

Innovation examples beyond customer support

Organizations have used AI for:

- Personalized advertising and campaign optimization
- Faster product development through pattern recognition on usage and feedback data
- Supply chain forecasting and intelligent routing to reduce bottlenecks and free working capital (Anchia, 2024; Dreyfus, 2025)

Across these examples, a consistent pattern appears: the strongest innovators **blend creative use cases with disciplined measurement**.

What these case studies teach, in plain terms

If you want a single sentence summary, it is this:

AI transformation succeeds when organizations treat AI as an operating capability, not a technology feature.

Across sectors, the leaders who win tend to do five things well:

- Define the business problem clearly and measure outcomes.
- Build trustworthy data foundations and governance.
- Integrate AI into workflows where decisions happen.
- Invest in adoption through training, champions, and change management.
- Monitor, learn, and improve continuously, with clear accountability.

Wrapping up

These case studies show real wins and real failures across retail, healthcare, manufacturing, and financial services. The lesson is not that AI always works or that AI often fails.

The lesson is that execution determines outcome.

When organizations invest in data quality, align AI with measurable goals, build trust through transparency and human oversight, and benchmark performance continuously, AI becomes a durable advantage. When they chase hype, skip governance, or ignore adoption, AI becomes expensive disappointment.

Bridge to the next chapter

In the next chapter, we will shift from case studies to leadership execution: the mindsets, decision frameworks, and practical habits that help business leaders and enterprise architects sustain AI-driven transformation over the long term, even as technology and market expectations continue to evolve.

> Success leaves clues. Failure leaves warnings.
> Learn from both.

CHAPTER 10

Leading with AI: Skills for Tomorrow's Leaders

The boardroom moment that changed the conversation

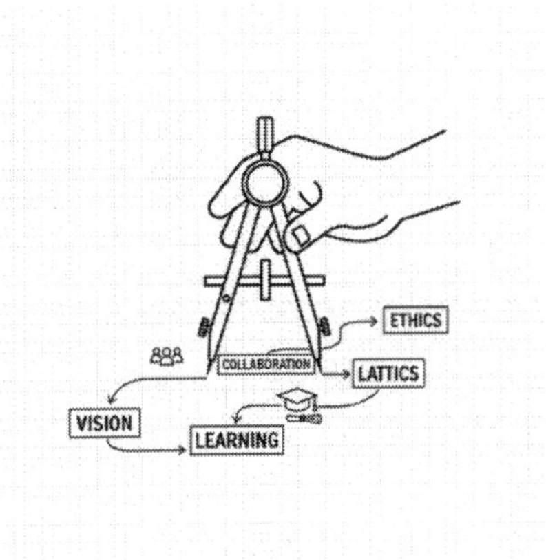

"Are we ready to lead with artificial intelligence, or are we just along for the ride?"

That question came up unexpectedly in a boardroom discussion not long ago.

The room went quiet for a beat, not because anyone doubted AI's impact, but because the question landed exactly where it needed to: leadership readiness.

One executive leaned forward and said, "We have pilots. We have a vendor roadmap. We have a data team. But do we have leaders who can guide the organization when the model is wrong, the data is messy, and the culture pushes back?"

That is the leadership challenge of the AI era. The rules are shifting quickly, and leaders are being asked to do something new: blend strategic clarity with enough technical understanding to ask sharp questions, create trust across diverse teams, and build resilience when change is constant.

This chapter focuses on the skills tomorrow's leaders need to make AI a real advantage, not a buzzword.

What AI leadership really means

Let's clear something up early. Leading with AI is not about knowing every algorithm, reading research papers at midnight, or learning to code.

It is about:

- Knowing enough to make good strategic decisions
- Asking the right questions at the right time
- Building teams that can deliver, learn, and improve
- Creating trust through ethics, transparency, and accountability
- Sustaining momentum when the organization gets tired

If AI is becoming part of how work gets done, then leadership must become part of how AI gets done.

Developing AI leadership competencies

Visionary thinking that stays grounded in reality

Visionary thinking in the AI era is not abstract dreaming. It is the ability to see how AI changes customer expectations, operating models, and competitive advantage, and then translate that into a practical direction your teams can execute.

A future-ready leader actively scans for signals:

- New AI-driven competitors entering the market
- Shifts in customer expectations, such as personalization becoming standard
- Regulatory changes triggered by automation and advanced analytics

A retail executive, for example, might recognize that AI-powered recommendations will soon be expected, not optional. Instead of waiting for competitors to set the pace, the leader communicates a clear strategy where personalization becomes central, and teams understand how their work connects to that vision (Palomino, 2025).

A useful leadership habit here is to keep asking:

What will customers expect next year that they do not expect today?

Then build toward that future in measured steps.

Technical acumen without drowning in jargon

You do not need to become a data scientist, but you do need technical fluency. Think of it as "executive-level AI literacy."

That means understanding concepts like machine learning, automation, and natural language processing well enough to ask questions such as:

- What decision are we automating, and what decision still needs human judgment?
- What data does this depend on, and how reliable is that data?
- What could go wrong, and how will we detect it early?
- What does success look like in business terms, not just model accuracy?

A leader who can distinguish between predictive analytics and simple rules-based automation can spot when an AI solution is being oversold, or when a tool is acting like a bandage instead of enabling real transformation (Artificial Intelligence and Leadership Agility: Concepts and Their Application -

International Journal of Research and Innovation in Social Science, 2025).

This technical baseline reduces miscommunication, improves decision quality, and increases trust between business leaders and technical teams.

Collaboration that turns experts into a unified team

AI transformation is a team sport. It crosses boundaries by design: data, security, architecture, operations, legal, HR, product, and customer-facing teams all have a role.

Leaders who unlock collaboration do a few things consistently:

- They make it safe for non-technical voices to contribute
- They set shared goals that teams can rally around
- They reward cross-functional problem solving, not silo success
- They create simple rhythms for alignment, such as weekly reviews and shared dashboards

Cross-functional teams can be powerful in surprising ways. A finance group may describe forecasting pain, and a data science team might propose an approach that reveals patterns no one saw before.

When leaders treat this exchange as normal, not exceptional, innovation becomes repeatable (Palomino, 2025).

A practical question to ask your leadership team is:

Do we have shared ownership of outcomes, or do we still hand work over the wall?

Adaptability and resilience when outcomes are uncertain

AI projects rarely move in a straight line. Models drift. Data changes. Adoption varies by region. Regulations evolve. Sometimes a promising use case simply does not deliver.

Resilient leaders treat setbacks as learning signals, not failures. They gather the team, examine what happened, and adjust the approach. They normalize iteration. They protect morale while still holding accountability.

This mindset is tied to leadership agility: leaders who encourage honest feedback and treat missed targets as inputs for improvement build organizations that learn faster (Artificial Intelligence and Leadership Agility: Concepts and Their Application - International Journal of Research and Innovation in Social Science, 2025).

Adaptability also means shifting course when needed. If a regulatory change disrupts a planned rollout, or if early pilots show unexpected risks, strong leaders adjust quickly while staying anchored to long-term outcomes (Palomino, 2025).

Fostering a culture of continuous learning

Even the best leaders cannot carry transformation alone. AI capability becomes durable when the organization learns as a system.

Create regular, structured knowledge sharing

Monthly "AI showcase" sessions work well: teams present what they built, what worked, what did not, and what they would do differently.

The goal is not celebration. The goal is shared learning.

Mentorship across departments also accelerates adoption. Pair an operations leader who rolled out automation with a marketing team starting a similar effort. That transfer of lived experience prevents repeat mistakes.

Digital knowledge platforms that capture experiments and lessons learned build a searchable memory for the organization. Research and workplace reporting in 2024 highlights that these practices can accelerate innovation and increase engagement by spreading practical expertise (Bradford, 2024).

Train for baseline literacy and role-based depth

Training should match roles. Executives need strategy, governance, and risk fluency. Managers need workflow and adoption skills. Technical teams need deep engineering and monitoring practices. Frontline teams need hands-on guidance that fits daily work.

Some organizations use executive education programs to help leaders build AI strategy and ethical awareness without requiring coding expertise, such as Harvard's professional and executive development offerings (AI Programs at Harvard - Professional & Executive Development | Harvard DCE, 2024).

The key is to avoid relying on a small group of experts. You want distributed competence.

Make experimentation safe and normal

If people fear mistakes, innovation slows. Leaders can create sandbox environments where teams can test AI tools without risking customer disruption.

Internal hackathons, limited pilots, and structured experiments help teams learn quickly. The point is not to avoid failure. The point is to fail safely, learn fast, and share what you learned.

Research on AI adoption has linked structured support for risk-taking with stronger innovation and measurable improvements (Bradford, 2024).

Recognize learning, not just winning

Recognition systems signal what matters. When leaders highlight teams that shared lessons, mentored others, or improved processes through iteration, they build a culture where learning is valued.

Bradford (2024) notes that recognition and reward systems can correlate with higher participation in upskilling and faster adoption of digital tools (Bradford, 2024).

Building resilient teams and leading ethically with AI

Build teams that blend technical and non-technical strengths

The strongest AI teams include machine learning engineers, data analysts, domain experts, designers, and operational leaders. This mix prevents the classic failure where the model is technically strong but operationally impractical.

When projects shift from model building to user trials, design thinking and empathy become central. Diverse teams pivot faster because the necessary skills are already present.

Clarify roles and decision rights

Resilient teams have clear ownership. Who owns data quality? Who owns model performance? Who owns workflow integration? Who owns adoption? Who can pause a system if risk emerges?

Role clarity reduces friction and prevents hidden gaps.

Create psychological safety so risks surface early

Psychological safety is not a soft concept. It is a practical control mechanism.

If a junior engineer sees bias, data leakage risk, or a flawed assumption, they must feel safe raising it. Teams with psychological safety correct issues sooner and adapt faster, which is especially important in AI projects where small issues can scale quickly (Alami et al., 2024).

Leaders shape this through behavior: listening to quieter voices, rewarding candor, and treating problems as solvable instead of punishable.

Make ethical leadership part of daily work

AI raises ethical issues from the start: bias, transparency, privacy, and responsible data use. Leaders should treat ethical checks as part of the workflow, not a last-minute review.

For example, a loan classification model should be tested for fairness as rigorously as it is tested for accuracy. Leaders can establish guidelines and monitoring practices that track data quality, model decisions, and ethical risk signals (José Andrés Gómez Gandía et al., 2025).

Ethics becomes real when leaders bring it into everyday conversation:

- "How do we know this is fair across groups?"
- "What would we say if a regulator asked us to explain this outcome?"
- "What data are we collecting, and do we truly need it?"
- "What is our plan when the model fails?"

Leaders who model this mindset create teams that take responsibility seriously and sustain trust over time (José Andrés Gómez Gandía et al., 2025; Alami et al., 2024).

Practical questions tomorrow's leaders should ask weekly

If you want a simple leadership routine, use these questions:

- What business outcome are our AI initiatives accountable for this quarter?
- Where is adoption lagging, and what is causing it?
- What data quality or governance issues are creating risk?
- What did we learn this week, and how will we apply it?
- What ethical or compliance risks could become external issues?
- Are we building reusable capabilities, or isolated one-off solutions?
- Do our teams feel safe raising problems early?

These questions keep leadership connected to reality and keep AI work aligned with business value and trust.

Bringing it all together

Leading with AI requires a new leadership toolkit: visionary thinking grounded in business outcomes, technical fluency that supports better questions, collaboration across silos, resilience when the path is messy, and ethical leadership that protects trust.

When leaders invest in continuous learning, create safe experimentation, clarify roles, and build psychological safety, they create an environment where AI can thrive and evolve with the market.

The journey ahead demands commitment, but leaders who develop these skills will turn AI from uncertainty into sustained transformation.

You don't need to be technical to lead, but you must be literate.

Conclusion

You have reached the end of this book, but not the end of the work. If there is one idea to carry forward, it is this: AI-driven transformation succeeds when artificial intelligence and enterprise architecture operate as one connected discipline, guided by clear strategy, strong governance, and measurable outcomes.

We started with fundamentals so you could build a shared language across business and IT. From there, we moved into execution: building a business case, aligning priorities, designing scalable architectures, and managing adoption in ways that respect the human side of change.

We treated ethics, security, and governance as core requirements because trust is what makes AI scalable, and we focused on measurement because value that cannot be proven rarely lasts.

We also looked at real case studies and future trends to help you avoid hype, learn faster, and stay ready as the landscape evolves.

Here is the practical roadmap you now have:

- **Align AI to business outcomes** so every initiative has purpose, ownership, and clear value.
- **Build data and architecture foundations** that allow AI to scale reliably across teams and use cases.
- **Embed ethics, security, and governance** so decisions are explainable, compliant, and trusted.
- **Manage change deliberately** through communication, training, and cross-functional collaboration.

- **Measure impact and improve continuously** using KPIs and feedback loops tied to business results.

The next step is simple, and it is not easy: choose one meaningful use case, define success in business terms, put the right guardrails in place, and deliver a result you can measure.

Then repeat, scaling what works and retiring what does not.

AI is moving fast, but lasting transformation is built through discipline and leadership. Keep learning, keep your teams connected, and keep trust at the center of your decisions.

If you do that, you will not just adopt AI. You will build an organization that can evolve with it.

> Your next step matters more than your next tool.
> Lead the system, not just the project.

Appendix: Templates and Tools

The chapters in this book are designed to help you move from ideas to execution. This appendix gives you a set of practical, one-page tools you can use to start immediately, whether you are shaping an AI opportunity, building a business case, reviewing architecture, standing up governance, or measuring outcomes.

Each tool is intentionally lightweight so you can use it in a workshop, a steering committee meeting, or an architecture review without turning it into a paperwork exercise. Think of these as your quick-start versions.

Tool 1: AI Use Case Intake Template

AI Use Case Intake

Use this when: you want a fast, consistent way to assess new AI ideas before they become pilots.

Use case title:

Business owner:

Sponsor:

Date submitted:

Business unit:

Process or capability impacted:

1) Problem and outcome

- **Problem statement (what is happening today):**
- **Desired outcome (what will be better):**
- **Who benefits (customers, employees, partners):**

2) AI opportunity

- **AI pattern (choose one):** Automation | Prediction | Recommendation | Detection | Personalization | Knowledge assistant
- **Decision type:** Human decision support | Human-in-the-loop | Fully automated (with oversight)
- **Primary users:**
- **Expected frequency and volume:**

3) Data and integration

- **Key data sources:**
- **Data quality status:** Unknown | Mixed | Good | Verified
- **Sensitive data involved:** None | Personal | Confidential | Regulated
- **Systems impacted or integrated:**

4) Value and measurement

- **Value hypothesis (how this creates business value):**
- **KPIs to improve (max 3):**
- **Baseline values (if known):**
- **Target values:**

5) Risks, governance, and constraints

- **Key risks (bias, privacy, security, operational):**
- **Controls needed (approval, monitoring, human oversight):**
- **Constraints (time, budget, skills, vendor):**

6) Delivery approach

- **Proposed pilot scope:**
- **Estimated timeline:**
- **Dependencies:**
- **Decision needed:** Approve discovery | Approve pilot | Defer | Decline

Tool 2: Business Case and ROI Worksheet

AI Business Case and ROI Summary

Use this when: you need a clear, board-ready view of costs, benefits, and payback.

Initiative name:

Executive sponsor:

Owner:

Time horizon: 12 | 24 | 36 months

1) Business objective

- **Strategic goal supported:**
- **Current pain (baseline):**
- **Success definition:**

2) Cost estimate (all-in)

One-time costs

- Data and integration:
- Model build or vendor:
- Platform or tooling:
- Security and governance:
- Change management and training:
- Other:
  Total one-time:

Ongoing annual costs

- Licensing and hosting:
- Model monitoring and retraining:
- Support and maintenance:
- Other:
- **Total ongoing:**

3) Benefits estimate

Tangible benefits (annualized)

- Labor savings:
- Error reduction or rework avoided:
- Cycle time reduction:
- Revenue lift or retention:
- Risk reduction cost avoided:
- **Total tangible:**

Intangible benefits

- Customer experience improvement:
- Better decision-making speed or quality:
- Compliance confidence:
- Employee experience and capability uplift:

4) ROI and payback

- **Net annual benefit:**
- **Payback period (months):**
- **ROI % (time horizon):**
- **Break-even assumptions:**

5) Scenarios

- **Best case (key assumptions):**
- **Base case (key assumptions):**
- **Worst case (key assumptions):**

6) Recommendation

Proceed | Proceed with conditions | Pilot first | Defer

Conditions or next steps:

Tool 3: Architecture Review Checklist

AI Architecture Review Checklist

Use this when: you are validating if a solution can be scaled and operated safely.

A) Alignment and scope

- Business outcome and KPIs are defined and agreed
- Use case fits target operating model and customer journey
- Decision rights and accountability are clear

B) Data readiness

- Data sources identified, owned, and accessible
- Data quality assessed and improvement plan exists
- Data privacy classification completed
- Data lineage and retention requirements documented

C) Model and AI controls

- Model purpose, inputs, outputs documented
- Explainability requirements defined for the use case
- Bias testing plan exists and is owned
- Human oversight points defined for critical decisions
- Monitoring plan covers drift, performance, and error handling

D) Integration and architecture

- Integration approach defined (APIs, events, batch)
- Architecture supports resiliency and failure modes
- Latency and performance requirements defined
- Logging and observability included
- Deployment approach supports repeatability and rollback

E) Security and compliance

- Threat model completed for AI-specific risks
- Access controls and least privilege applied
- Secrets and keys managed securely
- Audit trails support compliance requirements
- Vendor risk assessed, if applicable

F) Operations and scale

- Support model defined (L1 to L3)
- Runbooks and escalation paths exist
- Cost model defined and monitored
- Retraining cadence and ownership agreed
- Exit strategy defined if the model is retired

Review outcome: Approved | Approved with conditions | Rework required

Conditions or actions:

Tool 4: Governance Charter Sample

AI Governance Charter (Starter)

Use this when: you need clear oversight without slowing delivery.

1) Purpose

Ensure AI initiatives are aligned to strategy, governed responsibly, and scaled safely with measurable outcomes.

2) Scope

Applies to all AI systems that:

- influence decisions, automate workflows, or interact with customers or employees
- process confidential, personal, or regulated data
- create material business, legal, or operational risk

3) Governance bodies and roles

- **Executive sponsor:** accountable for outcomes and funding
- **AI product owner:** accountable for value delivery and adoption
- **Enterprise architecture:** accountable for fit, patterns, integration, and scalability
- **Data governance lead:** accountable for data quality, access, lineage, privacy
- **Security and risk:** accountable for threat modeling, controls, incident response
- **Legal and compliance:** accountable for regulatory alignment and disclosures

4) Decision rights

- Approve use case intake and prioritization
- Approve data access and privacy classification
- Approve model release for production and major changes
- Define human oversight requirements for critical decisions
- Approve KPI measurement and benefits tracking

5) Required artifacts

Use case intake, business case summary, architecture review, risk assessment, monitoring plan, KPI dashboard.

6) Cadence and reporting

- Steering review: monthly
- Risk and compliance review: quarterly
- Model performance review: monthly or per release
- Incident reporting: within agreed SLA

Tool 5: KPI Dashboard Starter Set

AI KPI Dashboard Starter Set

Use this when: you need a consistent, measurable view of AI value, adoption, and risk.

1) Value outcomes (pick 3 to 5)

- Cost per transaction
- Cycle time reduction
- Revenue lift or conversion improvement
- Customer satisfaction or NPS shift
- Error rate or rework reduction
- Fraud, loss, or risk events reduced

2) Model performance

- Accuracy, precision, recall (as relevant)
- False positives and false negatives
- Drift indicators (data and concept drift)
- Model uptime and latency
- Retraining frequency and results

3) Adoption and usage

- Active users and usage frequency
- Task completion rates
- Human override rate
- Time saved per user or team
- Training completion and confidence score

4) Risk and governance

- Incidents and severity
- Compliance exceptions
- Privacy events or access violations

- Bias or fairness audit results
- Security findings and remediation status

5) Operating health

- Cost to run (monthly)
- Cost per inference or transaction
- Support tickets by type
- Release frequency and rollback events

Dashboard notes

- Show baselines, targets, and trends
- Review monthly, with a quarterly value confirmation
- Keep ownership clear for every metric

Glossary

Ability (ADKAR)

The point where a person can actually use the new tool or process in real work, not just understand it in training.

Accountability

Clear ownership for decisions, outcomes, and controls, including who is responsible when AI produces harm or error.

Adversarial Attack

A technique that manipulates an AI system's inputs or training data to cause incorrect or harmful outcomes.

Agile Enterprise Architecture

An approach to architecture that uses short delivery cycles, frequent feedback, and iterative design instead of long, rigid plans.

AI (Artificial Intelligence)

A set of techniques that enable machines to perform tasks that usually require human intelligence, such as pattern recognition, language understanding, and decision support.

AI Champions (Change Champions)

Trusted individuals who advocate for an AI initiative, help peers adopt it, and provide feedback to the program team.

AI Containment Rate

A customer support measure showing the share of inquiries resolved by AI without needing a human agent.

AI Governance

The policies, roles, decision rights, and controls that ensure AI is used safely, ethically, legally, and in alignment with business goals.

AI Literacy

A practical understanding of what AI can and cannot do, plus how to ask the right questions about risk, value, and limitations.

AI Orchestration

Coordinating models, services, data flows, and automation so AI works reliably across systems and processes.

AI-Driven Enterprise Architecture

Designing operating model, data, systems, and governance so AI can be deployed safely, scaled reliably, and tied directly to business outcomes.

Algorithm

A set of steps or rules a system follows to reach a result, such as a prediction or recommendation.

Architecture Development Method (ADM)

TOGAF's structured cycle for developing and governing enterprise architecture from vision through implementation.

Architecture Principles

Guidelines that steer design decisions, such as "reuse before buy" or "data is a shared asset."

Audit Trail

A record of actions taken, such as who accessed data, changed a model, or approved a decision.

Automation

Using technology to perform tasks with minimal human effort, often to reduce manual work and errors.

Awareness (ADKAR)

Understanding why a change is happening and why it matters.

Benchmarking

Comparing performance against your past results, peers, or industry standards to set targets and identify gaps.

Bias

Systematic unfairness in AI outcomes, often caused by skewed data, flawed assumptions, or uneven model performance across groups.

Bias Audit

A structured review of data and model outcomes to detect and reduce unfair or discriminatory behaviour.

Blockchain

A tamper-evident ledger that can improve traceability of data, transactions, or changes in shared environments.

Business Case

A structured justification for investment, including expected benefits, costs, risks, and how success will be measured.

Capability Model

A map of what the business must be able to do, independent of organization structure or systems, often used to guide investment decisions.

CCPA

California's privacy law that governs certain uses, disclosures, and rights related to personal information.

Change Management

The structured approach for helping people adopt new ways of working, including communication, training, and reinforcement.

Cloud-native Architecture

Designing systems for elasticity, resilience, and fast deployment using cloud patterns such as containers and managed services.

Compliance

Meeting legal, regulatory, and internal policy requirements, often including privacy, security, and auditability.

Consent Management

Capturing, storing, and enforcing what people have agreed to regarding use of their data.

Continuous Learning

An organizational habit of building skills over time through training, practice, and shared lessons learned.

Cost-benefit Analysis

Comparing total costs of an initiative with expected benefits to evaluate whether it is worth doing.

Cross-functional Team

A team that blends business, technical, operational, and risk perspectives to deliver outcomes end to end.

Data Governance

The decisions, standards, and accountability that define how data is collected, managed, secured, and used.

Data Pipeline

The end-to-end flow that moves data from sources through processing into storage and AI or analytics systems.

Data Quality

How accurate, complete, consistent, timely, and usable data is for a given purpose.

Data Steward

A role accountable for data definitions, quality, access, and lifecycle practices.

Deep Learning

A machine learning approach using multi-layer neural networks, often effective for images, audio, and language at scale.

Desire (ADKAR)

The motivation to support and participate in change.

Digital Transformation

Changing how an organization operates and delivers value by using digital technologies and modern ways of working.

Distributed Data

Data stored and processed across multiple locations or systems to improve performance, resilience, or local access.

Enterprise Architecture (EA)

A discipline that aligns strategy, operating model, information, systems, and technology standards to guide change and investment.

Explainable AI (XAI)

Methods that help humans understand why a model produced a result, especially in high-stakes decisions.

Fairness

A goal of ensuring AI outcomes do not unjustly disadvantage individuals or groups.

Feedback Loop

A cycle where results and user input are used to improve the model, process, or product.

GDPR

The European privacy regulation governing personal data processing, rights, and organizational responsibilities.

General AI

A theoretical form of AI with broad reasoning and transfer learning ability across domains similar to humans.

HIPAA

A US health privacy and security law governing protected health information in certain contexts.

Human-in-the-loop

A design where humans review, approve, or override AI outputs, especially when risk is high.

Internet of Things (IoT)

Networks of connected sensors and devices that generate real-time data about physical environments and assets.

Key Performance Indicators (KPIs)

Measures used to track progress toward goals, such as cycle time, cost per transaction, accuracy, or adoption.

Kotter's 8-Step Model

A structured change approach emphasizing urgency, coalition-building, vision, removing barriers, and reinforcing change.

Knowledge (ADKAR)

Knowing how to change, including training and understanding the new process or tool.

Load Balancing

Distributing workloads across resources so systems remain fast and reliable during spikes.

Machine Learning (ML)

Training a model to learn patterns from data and make predictions or classifications.

Measurement Strategy

A plan for how outcomes will be tracked, reported, and used to guide decisions.

Microservices

An architecture style that breaks a system into smaller, independently deployable services.

Model Card

A standard artifact describing what a model does, how it was trained, limitations, and intended use.

Model Retraining

Updating a model over time with new data to maintain accuracy and relevance.

Model-Driven Architecture (MDA)

Designing systems using formal models that guide implementation, often improving consistency and traceability.

Narrow AI

AI designed for a specific task, such as fraud detection or document processing, which is the most common form in business today.

Natural Language Processing (NLP)

Techniques that allow systems to interpret and generate human language, including classification, extraction, and conversational interfaces.

Observability

The ability to understand system behavior through logs, metrics, traces, and monitoring, including AI performance signals.

Operating Model

How a business delivers value, including roles, processes, decision rights, governance, and ways of working.

Penetration Testing

Simulated attacks against systems to identify security weaknesses before real attackers do.

Predictive Analytics

Using data and models to forecast likely outcomes, such as churn, demand, or failure risk.

Privacy-by-design

Embedding privacy controls into systems from the start, not adding them later.

Prosci ADKAR Model

A change model focused on individual adoption through Awareness, Desire, Knowledge, Ability, and Reinforcement.

Reference Architecture

A reusable blueprint that provides common patterns and guidance for designing solutions.

Reinforcement (ADKAR)

Actions that sustain change over time, such as recognition, updated metrics, and ongoing support.

Responsible AI

A practical approach to building and using AI with fairness, transparency, accountability, privacy, and safety.

Return on Investment (ROI)

A financial measure comparing benefits gained to costs incurred.

Role-based Access Control (RBAC)

Limiting system access based on a person's role so users only have the permissions they need.

Scenario Planning

Testing best-case, expected, and worst-case outcomes to plan budgets and manage uncertainty.

Security-by-design

Building security controls into architecture and development from the beginning, including monitoring and incident response planning.

SMART Objectives

Goals that are Specific, Measurable, Achievable, Relevant, and Time-bound.

Stakeholder Buy-in

Gaining support and commitment from people affected by or funding the initiative.

Superintelligent AI

A theoretical AI level that surpasses human intelligence across domains.

TOGAF

A widely used enterprise architecture framework that includes the Architecture Development Method and governance practices.

Transparency

Providing clarity about how AI is used, what data it relies on, and how decisions are made or explained.

Total Cost of Ownership (TCO)

The full cost of a system over time, including build, licensing, maintenance, operations, and change impacts.

Value Realization

Turning planned benefits into real outcomes through adoption, measurement, and continuous improvement.

References

Adamec, V. (2024, July 12). Where Is Your Company on the AI Journey? An AI Adoption Roadmap for Business Leaders. *Gaussalgo.com*; Gauss Algorithmic. https://www.gaussalgo.com/knowledge-base/ai-adoption-roadmap-for-business-leaders

AI Case Studies in Healthcare, Finance, Retail & More. (2020, March 16). *Webmobtech.com*. https://webmobtech.com/blog/ai-in-business-case-studies/

AI Programs at Harvard - Professional & Executive Development | Harvard DCE. (2024, September 11). *Professional & Executive Development | Harvard DCE*. https://professional.dce.harvard.edu/ai-courses/

Alami, A., Zahedi, M., & Krancher, O. (2024, July 25). The role of psychological safety in promoting software quality in agile teams. *Empirical Software Engineering*; Springer Science+Business Media. https://doi.org/10.1007/s10664-024-10512-1

Ali, M., Khan, T. I., Khattak, M. N., & ŞENER, İ. (2024, August 1). Synergizing AI and Business: Maximizing Innovation, Creativity, Decision Precision, and Operational Efficiency in High-Tech Enterprises. *Journal of Open Innovation Technology Market and Complexity*; Springer Science+Business Media. https://doi.org/10.1016/j.joitmc.2024.100352

Anchia, C. (2024, April 11). Measuring Success: Key Metrics and KPIs for AI Initiatives. *Acacia Advisors*. https://chooseacacia.com/measuring-success-key-metrics-and-kpis-for-ai-initiatives/

Artificial Intelligence and Leadership Agility: Concepts and their Application - International Journal of Research and Innovation in Social Science. (2025). *International Journal of Research and Innovation in Social Science*. https://rsisinternational.org/journals/ijriss/articles/artificial-intelligence-and-leadership-agility-concepts-and-their-application/

Auday Al-Dulaimy, Jansen, M., Johansson, B., Trivedi, A., Alexandru Iosup, Ashjaei, M., Galletta, A., Dragi Kimovski, Prodan, R., Konstantinos Tserpes, Kousiouris, G., Giannakos, C., Ivona Brandic, Ali, N., Bondi, A. B., & Papadopoulos, A. V. (2024, October 1). The computing continuum: From IoT to the cloud. *Internet of Things*; Elsevier BV. https://doi.org/10.1016/j.iot.2024.101272

Babina, T., Fedyk, A., He, A., & Hodson, J. (2024, January 1). Artificial intelligence, Firm growth, and Product Innovation. *Journal of Financial Economics*; Elsevier BV. https://doi.org/10.1016/j.jfineco.2023.103745

Bhardwaj, C. (2025, September 19). How to Assess Your Organization's AI Maturity Level: A Deep Dive into Key Frameworks and Implementation Strategies. *Appinventiv*. https://appinventiv.com/blog/ai-maturity-assessment/

Bradford, N. (2024). 7 Ways 2024 Was a Year of AI Experimentation and Learning. *Shrm.org*. https://www.shrm.org/topics-tools/flagships/ai-hi/year-of-ai-experimentation-learning

Brusati, I. (2024). ADKAR vs Kotter: Which Change Model Should You Choose? *Prosci.com*; Prosci Inc. https://www.prosci.com/blog/adkar-vs-kotter

Business, A. (2022). Using AI to Empower Cross-Functional Teams. *Agilebusiness.org*. https://www.agilebusiness.org/resource/using-ai-to-empower-cross-functional-teams.html

DigitalDefynd, T. (2024, May 2). 5 AI in Finance Case Studies [2024]. *DigitalDefynd*. https://digitaldefynd.com/IQ/ai-in-finance-case-studies/

Dilmegani, C. (2018, February 16). 100+ AI Use Cases & Applications in 2021: In-Depth Guide. *Research.aimultiple.com*. https://research.aimultiple.com/ai-usecases/

DILAN, T. (2025, February 21). Dilan Consulting Group. *DILAN Consulting*. https://dilanconsulting.com/the-leadership-playbook-six-essential-skills-for-2025-and-beyond/

Dreyfus, G. (2025, April 29). AI Benchmarking Best Practices: A Framework for CX Leaders. *Quiq*. https://quiq.com/blog/ai-benchmarking-best-practices/

DX. (2025). Getdx.com. https://getdx.com/blog/ai-roi-enterprise/

Enterprise IT Services. (2025). JanBask. https://www.janbask.com/enterprise-it-services

Erickson, J. (2024, June 21). The Role and Benefits of AI in Cloud Computing. *Oracle.com*; Oracle. https://www.oracle.com/artificial-intelligence/ai-cloud-computing/

GAN Integrity. (2024, July 23). AI Governance, A Critical Framework for Organizations | GAN Integrity. *Ganintegrity.com*; GAN Integrity. https://www.ganintegrity.com/resources/blog/ai-governance/

Ghosh, U. K. (2025, January 17). Transformative AI Applications in Business Decision-Making. *Advances in Logistics, Operations, and Management Science Book Series*; Routledge. https://doi.org/10.4018/979-8-3373-1687-1.ch001

Gibson, K. (2024, September 10). AI-Driven Business Models: 4 Characteristics | HBS Online. *Business Insights Blog*. https://online.hbs.edu/blog/post/ai-driven-business-models

Gibson, K. (2024, September 25). The Role of Artificial Intelligence in Digital Transformation. *Business Insights Blog*. https://online.hbs.edu/blog/post/ai-digital-transformation

Harris, N. (2023, September 28). How to Measure Marketing Effectiveness by Setting Goals. *Www.npws.net*. https://www.npws.net/blog/measure-marketing-effectiveness-setting-goals

Jada, I., & Mayayise, T. O. (2023). The impact of artificial intelligence on organisational cyber security: An outcome of a systematic literature review. *Data and Information Management*; De Gruyter; Sciencedirect. https://doi.org/10.1016/j.dim.2023.100063

Jager, E. (2025, September 22). Architecting with AI – EA Voices. *Eavoices.com*. https://eavoices.com/2025/09/22/architecting-with-ai/

Jacob. (2025, August 8). Measuring AI Performance in Competitive Markets: 7 Metrics That Matter (2025) 🚀. *ChatBench*. https://www.chatbench.org/measuring-ai-performance-in-competitive-markets/

José Andrés Gómez Gandía, Antonio, & Teresa. (2025, June 11). Knowledge and artificial intelligence on employee behaviour advancing safe and respectful workplace. *Journal of*

Innovation & Knowledge; Elsevier BV. https://doi.org/10.1016/j.jik.2025.100750

Kassa, B. Y., & Worku, E. K. (2025, January). The Impact of Artificial Intelligence on Organizational Performance: The Mediating Role of Employee Productivity. *Journal of Open Innovation: Technology, Market, and Complexity*; Elsevier BV. https://doi.org/10.1016/j.joitmc.2025.100474

Marsh, B. (2024, April 19). Unlocking Growth: A Comprehensive Guide to AI Adoption for Businesses. *RTS Labs*. https://rtslabs.com/ai-adoption-for-business

Michał. (2024, September 27). Case Studies in Transforming AI Process Automation Across Sectors - FutureCode IT Consulting. *FutureCode IT Consulting*. https://future-code.dev/en/blog/case-studies-in-transforming-ai-process-automation-across-sectors/

Mucci, T., & Stryker, C. (2024, October 10). AI Governance. *Ibm.com*. https://www.ibm.com/think/topics/ai-governance

Navigating the Future of Leadership in Business Landscapes: Key Trends for 2025 and Beyond. (2025). *Cui.edu*. https://www.cui.edu/townsend/articles/post/navigating-the-future-of-leadership-in-business-landscapes-key-trends-for-2025-and-beyond

Papagiannidis, E., Mikalef, P., & Conboy, K. (2025, January 5). Responsible artificial intelligence governance: A review and research framework. *The Journal of Strategic Information Systems*; Elsevier BV. https://doi.org/10.1016/j.jsis.2024.101885

Palomino, I. (2025, February 22). Top 10 Qualities of Effective AI Leaders in the Tech Era. *Peoplekult*.

https://www.peoplekult.com/post/top-10-qualities-of-effective-ai-leaders-in-the-tech-era

Prosci. (2024, September 2). Kotter's Change Management Theory Explanation and Applications. *Prosci*; Prosci Inc. https://www.prosci.com/es/blog/recursos-humanos-clave-implementar-gestion-cambio-empresa

Proving ROI - Measuring the Business Value of Enterprise AI - agility at scale. (2025, April 4). *Agility at Scale*. https://agility-at-scale.com/implementing/roi-of-enterprise-ai/

Rajagopalan, R. (2020, November 25). Artificial Intelligence in Business: 10 Notable Examples. *University of San Diego*. https://onlinedegrees.sandiego.edu/artificial-intelligence-business/

Rashid, A. B., & Kausik, A. K. (2024). AI Revolutionizing Industries Worldwide: a Comprehensive Overview of Its Diverse Applications. *Hybrid Advances*; Elsevier BV. https://doi.org/10.1016/j.hybadv.2024.100277

Schrage, M., & Kiron, D. (2025, July 15). Winning With Intelligent Choice Architectures. *MIT Sloan Management Review*. https://sloanreview.mit.edu/projects/winning-with-intelligent-choice-architectures/

Soulami, M., Benchekroun, S., & Galiulina, A. (2024, November 14). Exploring How AI Adoption in the Workplace Affects employees: a Bibliometric and Systematic Review. *Frontiers in Artificial Intelligence*; Frontiers Media SA. https://doi.org/10.3389/frai.2024.1473872

Stepan Ilyin. (2025, July 17). AI Security: Risks, Frameworks, and Best Practices. *Wallarm.com*; Wallarm.

https://www.wallarm.com/what/ai-security-risks-frameworks-and-best-practices

Swift, P. (2025, June 10). How AI-Powered Change Communication Powers Transformation - Cerkl Broadcast. *Cerkl Broadcast*. https://cerkl.com/blog/ai-in-change-communication/

Takyar, A. (2023, December 7). AI in business management. *LeewayHertz - AI Development Company*. https://www.leewayhertz.com/ai-in-business-management/

Team, D. (2024, June 3). Cross-Functional Teams for AI Success: Guide. *Dialzara.com*; Dialzara. https://dialzara.com/blog/cross-functional-teams-for-ai-success-guide

The Role of AI in HR Continues to Expand. (2025). *Shrm.org*. https://www.shrm.org/topics-tools/research/2025-talent-trends/ai-in-hr

Tweed, L. (2025, April 16). AI for Enterprise Architecture: Automating Manual Workflows at Scale. *IEEE Computer Society*. https://www.computer.org/publications/tech-news/trends/ai-enterprise-arch/

Understanding AI Strategy: Key Principles for Success in Today's Digital Era. (2024). *Thestrategyinstitute.org*. https://www.thestrategyinstitute.org/insights/understanding-ai-strategy-key-principles-for-success-in-todays-digital-era

Vuković, D. B., Dekpo-Adza, S., & Matović, S. (2025, April 22). AI integration in financial services: a systematic review of trends and regulatory challenges. *Humanities and Social Sciences Communications*; Springer Science and Business Media LLC. https://doi.org/10.1057/s41599-025-04850-8

2025 Volume 5 How to Measure and Prove the Value of Your AI Investments. (2025). *ISACA*. https://www.isaca.org/resources/news-and-trends/newsletters/atisaca/2025/volume-5/how-to-measure-and-prove-the-value-of-your-ai-investments

(2025). *Journalofsocialsciences.org*. https://journalofsocialsciences.org/article/the-impact-of-artificial-intelligence-on-learning-and-development-case-studies-in-companies-of-dif/

Made in the USA
Las Vegas, NV
17 February 2026

42117789R00107